Lecture Notes in Economics and Mathematical Systems

Managing Editors: M. Beckmann and W. Krelle

322

Tryphon Kollintzas (Ed.)

The Rational Expectations Equilibrium Inventory Model

Theory and Applications

Springer-Verlag

New York Berlin Heidelberg London Paris Tokyo

Editorial Board

H. Albach M. Beckmann (Managing Editor) P. Dhrymes
G. Fandel G. Feichtinger J. Green W. Hildenbrand W. Krelle (Managing Editor)
H.P. Künzi K. Ritter R. Sato U. Schittko P. Schönfeld R. Selten

Managing Editors

Prof. Dr. M. Beckmann
Brown University
Providence, RI 02912, USA

Prof. Dr. W. Krelle
Institut für Gesellschafts- und Wirtschaftswissenschaften
der Universität Bonn
Adenauerallee 24-42, D-5300 Bonn, FRG

Editor

Professor Tryphon Kollintzas
Department of Economics, University of Pittsburgh
Pittsburgh, PA 15260, USA

ISBN 0-387-96940-3 Springer-Verlag New York Berlin Heidelberg
ISBN 3-540-96940-3 Springer-Verlag Berlin Heidelberg New York

This work is subject to copyright. All rights are reserved, whether the whole or part of the material is concerned, specifically the rights of translation, reprinting, re-use of illustrations, recitation, broadcasting, reproduction on microfilms or in other ways, and storage in data banks. Duplication of this publication or parts thereof is only permitted under the provisions of the German Copyright Law of September 9, 1965, in its version of June 24, 1985, and a copyright fee must always be paid. Violations fall under the prosecution act of the German Copyright Law.

© Springer-Verlag New York, Inc., 1989
Printed in Germany

Printing and binding: Druckhaus Beltz, Hemsbach/Bergstr.
2847/3140-543210

To my mother Florentia

ACKNOWLEDGEMENTS

This volume would not have been possible without the International Society for Inventory Research. The Society gave me the opportunity to organize two sessions on aggregate inventory behavior in its First North American Meeting at Wesleyan University, June 1987. The sessions were as follows:

Session I:

Chairman:	Robert M. Solow
Paper:	Sophia P. Dimelis and Tryphon Kollintzas: "A Linear Rational Expectations Equilibrium Inventory Model for the American Petroleum Industry."
Discussants:	Alan S. Blinder
	Martin Eichenbaum
Paper:	Jeffrey A. Miron and Stephen P. Zeldes: "Seasonality, Cost Shocks, and the Production Smoothing Model of Inventories."
Discussants:	James A. Kahn

Session II:

Chairman:	Alan S. Blinder
Paper:	Lawrence J. Christiano and Martin Eichenbaum: "Temporal Aggregation and Structural Inference in the Inventory Stock Adjustment Model."
Discussants:	John C. Haltiwanger
	Robert M. Solow
Paper:	Kenneth D. West: "Unfilled Orders and Production Smoothing."
Discussants:	Louis Maccini
	Jeffrey A. Miron

The last four chapters of this volume contain the papers given in these sessions. I am therefore most grateful to the participants in these sessions as well as the organizers and especially Attila Chikan and Michael Lovell. I am likewise grateful to _Econometrica_ for giving me permission to reprint the paper by Miron and Zeldes. Last but not least I am grateful to Karen Troiani for expert and tireless word processing. She also helped me proof read the material. My quest to finish the book in time for publication in 1988 became hers and for this I could never repay her.

Over the last two years, during which this volume was put together, I have visited the Athens School of Economics and Business, the Federal Reserve Bank of Minneapolis and the University of Minnesota. Also, during these years I have received support from the National Science Foundation and the University of Pittsburgh.

TK

PREFACE

This volume consists of six essays that develop and/or apply "rational expectations equilibrium inventory models" to study the time series behavior of production, sales, prices, and inventories at the industry level. By "rational expectations equilibrium inventory model" I mean the extension of the inventory model of Holt, Modigliani, Muth, and Simon (1960) to account for: (i) discounting, (ii) infinite horizon planning, (iii) observed and unobserved by the "econometrician" stochastic shocks in the production, factor adjustment, storage, and backorders management processes of firms, as well as in the demand they face for their products; and (iv) rational expectations. As is well known according to the Holt et al. model firms hold inventories in order to: (a) smooth production, (b) smooth production changes, and (c) avoid stockouts. Following the work of Zabel (1972), Maccini (1976), Reagan (1982), and Reagan and Weitzman (1982), Blinder (1982) laid the foundations of the rational expectations equilibrium inventory model. To the three reasons for holding inventories in the model of Holt et al. was added (d) optimal pricing. Moreover, the popular "accelerator" or "partial adjustment" inventory behavior equation of Lovell (1961) received its microfoundations and thus overcame the "Lucas critique of econometric modelling." More recently, there has been a number of studies that seek to test the rational expectations equilibrium inventory model and, in particular, to examine whether this model is consistent with the stylized facts identified by Blinder (1986): (α) the variance of production exceeds that of sales, (β) the covariance between sales and inventory investment is positive, and (γ) "speed of adjustment" estimates from partial adjustment investment equations are unrealistically slow.

In Chapter III of this volume, Christiano and Eichenbaum investigate whether temporal aggregation bias can account for (γ). Temporal aggregation bias occurs when the economic agents' decision interval is smaller than the data sampling interval. First they develop a continuous time linear rational expectations equilibrium inventory model and then they develop its discrete time counterpart. By linear rational expectations equilibrium model I mean a rational expectations equilibrium model where the objective functions of economic agents are quadratic, the transition constraints they face are linear, and conditional expectations are taken to be the corresponding minimum mean square error predictors. Christiano and Eichenbaum estimate both continuous and discrete time models and after a series of tests conclude that temporal aggregation bias may significantly distort "speed of adjustment" estimates.

In Chapter IV Dimelis and Kollintzas develop a linear rational expectations equilibrium inventory model to characterize the behavior of refineries in the American petroleum industry and they couple this with a somewhat similar model of exhaustible resource supply to characterize the behavior of crude petroleum producers in this industry. Their empirical results suggest (a) despite evidence of significant technology shocks (iii) and they find evidence contrary to (γ). Further, their results suggest that crude and refined petroleum product inventories behave quite differently. So that serious aggregation biases may arise when those inventory series are aggregated.

In Chapter V Miron and Zeldes investigate whether the apparent inability of models that incorporate only reason (a) for holding inventories to account for (α) is due to failure to take into consideration observed and unobserved stochastic shocks to the production technology of firms and/or failure to estimate the model on seasonally unadjusted data. Shocks of this kind create a motive to hold inventories in order to smooth sales rather than production. Seasonal fluctuations in sales are the kind of fluctuations that firms should most easily identify and thus accommodate by smoothing production. After a number of ingenious tests they conclude that shocks to the production technology and failure to estimate the model on seasonally unadjusted data cannot account for this model's inability to explain (α).

In Chapter VI, West investigates whether the apparent inability of linear rational expectations equilibrium inventory models that do not incorporate reason (d) for holding inventories and ignore the technology shocks in (iii), to account for (α) and (β) is due to ignoring backlogs. That is, although in these models the nonnegativity constraint on inventories is ignored and backlogs are considered to be negative inventories, in practice, inventories are only physical inventories. His results suggest that when inventories are physical inventories minus backlogs (α) and (β) are reversed and the models he addressed can account for that reversal.

The first two chapters are more theoretical in nature. In Chapter II, Aiyagari, Eckstein, and Eichenbaum develop a version of the rational expectations equilibrium inventory model which incorporates the above mentioned nonnegativity constraint on inventories. They examine the cases of a competitive industry and of monopoly. They find that the nonnegativity constraint will be occasionally binding under both market structures depending on the relationship between a "reservation" and the actual price. Further, they provide a linear rational expectations equilibrium example whereby they calculate the equilibria under both market structures and they investigate their comparative dynamics properties. In Chapter I, which is meant to serve as an introduction to the other papers in this volume, I develop a linear rational expectations equilibrium inventory model. The necessary conditions and the sufficient conditions for its solution are derived. Further, the stability, cycling, and comparative dynamics properties of the solution are investigated.

This volume should be primarily of interest to researchers in business cycle phenomena and applied industrial organization. However, it should also be useful to students in applied dynamic economics and econometrics., in general.

References

Blinder, Alan S. (1982): "Inventories and Sticky Prices: More on the Microfoundations of Macroeconomics," American Economic Review, 72, 334-348.

Blinder, Alan S. (1986): "Can the Production Smoothing Model of Inventory Behavior be Saved?", Quarterly Journal of Economics, 101, 431-454.

Holt, Charles C., Franco Modigliani, John F. Muth, and Herbert A. Simon (1960): Planning Production, Inventories and Work Force, Englewood Cliffs, N.J.: Prentice Hall.

Lovell, Michael C. (1961): "Manufacturers' Inventories Sales Expectations, and the Accelerator Principle, Econometrica, 29, 293-314.

Maccini, Louis J. (1976): "An Aggregate Dynamic Model of Short-Run Price and Output Behavior," Quarterly Journal of Economics, 90, 177-196.

Reagan, Patricia B. (1982): "Inventories and Price Behavior," Review of Economic Studies, 49, 137-142.

Reagan, Patricia B., and Martin L. Weitzman (1982): "Price and Quantity Adjustment by the Competitive Industry," Journal of Economic Theory, 15, 440-420.

Zabel, Edward (1972): "Multiperiod Monopoly Under Uncertainty," Journal of Economic Theory, 5, 524-536.

TABLE OF CONTENTS

Chapter I:

    The Linear Rational Expectations Equilibrium Inventory Model: An Introduction, by **Tryphon Kollintzas**... 1

Chapter II:

    Inventories and Price Fluctuations under Perfect Competition and Monopoly, by **S. Rao Aiyagari, Zvi Eckstein, and Martin Eichenbaum**...................... 34

Chapter III:

    Temporal Aggregation and the Stock Adjustment Model of Inventories, by **Lawrence J. Christiano and Martin Eichenbaum**...................................... 70

Chapter IV:

    A Linear Rational Expectations Equilibrium Model for the American Petroleum Industry, by **Sophia P. Dimelis and Tryphon Kollintzas**.......................... 110

Chapter V:

    Seasonality, Cost Shocks, and the Production Smoothing Model of Inventories, by **Jeffrey A. Miron and Stephen P. Zeldes**... 199

Chapter VI:

    Order Backlogs and Production Smoothing, by **Kenneth D. West**.. 246

CHAPTER I

THE LINEAR RATIONAL EXPECTATIONS EQUILIBRIUM INVENTORY MODEL:
AN INTRODUCTION[1]

Tryphon Kollintzas
University of Pittsburgh

Abstract.

This paper intends to serve as an introduction to the other papers in this volume. It develops a linear rational expectations equilibrium version of the Holt et al (1960) inventory model. The necessary conditions and the sufficient conditions for a solution are derived and the stability, cycling, and comparative dynamics properties of the solution are investigated.

1. **Introduction**

Over the last ten years an extensive literature on the time series behavior of production, sales, inventories, and prices at the industry as well as the economy level has been developed. This literature is of central importance to macroeconomics for its objective is no less than the understanding of the business cycle and the proper role of stabilization policy. The theoretical model underlying most of the studies in this literature is what I shall call the "linear rational expectations equilibrium inventory model". Briefly, this is the extension of the model of Holt, Modigliani, Muth, and Simon (1960) to account for: (i) discounting, (ii) infinite horizon planning, (iii) observed and unobserved by the "econometrician" stochastic shocks in the production, factor adjustment, storage, and backorders management processes of firms, as well as the demand they face for their products; and (iv) rational expectations (i.e., firms forming their expectations about variables of interest to them based on the objective laws of motion of the physical environment and the actual strategies of other

---

[1] I am grateful to Kenneth D. West for valuable discussions and comments. The usual proviso applies.

economic agents they interact with). Versions of this model have been used in Blinder (1982, 1986), Blanchard (1983), Eichenbaum (1983, 1984), West (1986) and many more other studies.[2]

Despite its popularity, however, this model remains somewhat of a mystery when it comes to things other than the Euler Condition for its solution. This is primarily because most authors are interested in empirical applications whereby the Euler Condition is unfortunately all that is tested.[3] In many cases the Transversality Condition (or some other stability condition, that is used to distinguish from the infinitely many solutions of the Euler Condition, the one that is the solution to the equilibrium problem) is ignored. Even when the Transversality Condition is mentioned, it is, more often than not, stated incorrectly. On the contrary, most authors set their models under strong curvature restrictions of the kind that ensure that if a solution to the Euler Condition that satisfies the Transversality Condition exists it is the solution to the equilibrium problem. This is, of course, unnecessary in empirical applications. Moreover, these conditions are too restrictive. For example, the rate at which production costs increase does not have to be strictly positive if the rate at which adjustment costs increase is assumed to be strictly positive. Neither the rate at which inventory holding costs increase has to be positive if the rate at which backlog costs increase is positive.

Even more importantly, except in the simplest cases, the dynamic properties of the equilibrium have not to my knowledge been characterized other than by simulations or impulse response analyses.[4] But these analyses do not separate the deterministic from the stochastic components of the solution. The deterministic behavior of

---

[2] This introduction is not intended to be a review of the literature. Readers interested in a bibliography of studies that develop and/or use the linear rational expectations equilibrium inventory model should consult the references in the above mentioned studies, the studies in the subsequent chapters of this book, and the recent papers by West (1987), Ramey (1987), and Eichenbaum (1988).

[3] The exception that proves the rule is Blinder (1982, 1986).

[4] Blinder (1982, 1986) ignores adjustment costs and backlog costs that we will find to be necessary for cycling in the case of profit maximization.

the equilibrium (i.e., the behavior of the equilibrium when all exogenous variables are set equal to their means) is crucial for such matters as growth, endogenous cycling, and periodic motion. If, for example, deterministic cycles are possible, excluding them before testing may bias the estimates of magnitude of the supply and/or demand shocks that usually define the stochastic part of the equilibrium. Answers to questions like: (i) Why deterministic cycles happen? and (ii) How discount rates or other parameters effect these cycles? have, obviously, major implications for business cycle theory and stabilization policy.

The purpose of this paper is to examine the necessary and sufficient conditions for a solution to the linear rational expectations equilibrium model and to characterize the deterministic dynamic properties of the equilibrium. The approach taken is pedagodical in nature. It is hoped that the readers of this paper would acquire the basic knowledge necessary to handle the more technically demanding papers that follow.

The remainder of this paper is organized as follows. Section 2 sets the model. Section 3 develops the necessary and sufficient conditions for a solution and obtains that solution. Section 4 characterizes the stability, periodicity, and cycling properties of the solution. Finally, Section 5 concludes.

2. A Simple Linear Rational Expectations Equilibrium Model

Consider an industry that consists of a fixed number of identical firms, N, producing a single storable, homogeneous, product. At the beginning of any period τ, $\tau \varepsilon \; N_+ \equiv \{0, 1, \ldots\}$, the representative firm in this industry seeks a "contingency plan" for its production and sales in all future preiods, $\{q_t, s_t\}_{t=\tau}^{\infty}$, so as to maximize its "expected real present value":[5]

$$E_\tau \sum_{t=\tau}^{\infty} \beta^{t-\tau} \Pi(s_t, q_t, i_t, q_{t-1}, \xi_t) \tag{2.1}$$

---

[5] These terms are rigorously defined later.

where E denotes the mathematical expectations operator and $E_\tau(\cdot) = E(\cdot \mid \Omega_\tau)$ denotes that expectations are conditioned on the firm's information at the beginning of period τ, Ω_τ, which will be defined later; $\beta \epsilon (0, 1)$ is the real discount factor of the firm in all periods; and

$$\Pi(s_t, q_t, i_t, q_{t-1}, \xi_t) = p_t s_t - [a\theta_t q_t + 1/2\, bq_t^2 +$$
$$1/2\, c(q_t - q_{t-1})^2] - [d\theta_t i_t + 1/2\, ei_t^2 + 1/2\, f(i_t - gs_t)^2] \quad (2.2)$$

denotes the firm's real profits in period t. Here, p_t is the relative price of the firm's product in period t; i_t is the net inventory (i.e., inventory less unfilled orders) of that product at the beginning of period t; $\theta_t \in (\underline{\theta}, \overline{\theta}) \subseteq R$ is a stochastic parameter that incorporates the influence of random production shocks and is the second element of the vector ξ_t; and a,...,g are constants.

The term inside the first bracket represents production and adjustment costs. The term inside the second bracket in (2.2) represents inventory and backlog costs. The justification for all these costs is given in Holt et al. The basic difference between the costs considered there and (2.2) is that the later includes the stochastic parameter θ_t.[6] Most seminal studies of inventory behavior over the last ten years assume a cost structure that is similar but more restrictive than the one assumed here. See Table 1, below. No a priori restrictions have been imposed on the constant parameters of (2.2) so that restrictions may be imposed as required by theory.

The firm's optimization problem is subject to the following constraints:

$$i_{t+1} = (1-\gamma) i_t + q_t - s_t \quad (2.3)$$

$$(q_t, s_t) \in R_+ \times R_+ \quad (2.4)$$

---

[6] Production, adjustment, and inventory holding costs are allowed to interact in Eichenbaum (1983, 1984). The consequences of this will be discussed in Sections 4 and 5.

$$(i_\tau, q_{\tau-1}) \in R \times R_+ \tag{2.5}$$

where $\gamma \in [0,1)$ is the inventory depreciation rate in all periods.

The aggregate demand curve facing this industry is given by:

$$S_t = \frac{\eta_t}{h} - \frac{1}{h} p_t \tag{2.6}$$

where $S_t = Ns_t$, $\eta_t \in (\underline{\eta}, \bar{\eta}) \subset R_+$ is a stochastic parameter that incorporates the influence of random demand shocks and is the first element of the vector ξ_t. Formally, we take $\frac{\eta_t}{Nh} \to s_t$ as $h \to \infty$. In this case the firm's sales are exogenously determined and the firm's problem is to minimize expected discounted future stream of real costs as in the studies by Holt et al. and Blanchard. This illustrates the second difference between the model in Holt et al. and the present model. That is, the latter allows for sales to be determined by the firms' optimizing behavior. Also, we take the case where $h \to 0$ to represent the situation where $p_t = \eta_t$. That is when the aggregate demand is horizontal.

The stochastic process $\{\xi_t: t \in N_+\}$, $\xi_t = (\eta_t, \theta_t)'$ characterizes the industry's stochastic environment. The transition of that process is characterized by:

$$\xi_{t+1} = A \xi_t + \varepsilon_{t+1} \tag{2.7}$$

$$\xi_\tau \quad \text{given} \tag{2.8}$$

where the eigenvalues of A have modulus less than $\beta^{-1/2}$ and $\{\varepsilon_t: t \in N_+\}$ is a vector white noise process with fixed probability distribution $F(\varepsilon)$, $\varepsilon \in (\underline{\varepsilon}, \bar{\varepsilon}) \in R^2$.[7]

The restriction on A ensures that $\{\xi_t: t \in N_+\}$ does not "grow" too fast. In particular, they imply that $\{\xi_t: t \in N_+\}$ belongs to the

---

[7] The results of this paper can be easily extended to the case where $\{\xi_t : t \in N_+\}$ is an ARMA process. See, Kollintzas and Geerts (1984) for an extension of the generalized Wiener-Kolmogorov prediction formula of Hansen and Sargent (1980) to the case of ARMA processes.

family of expected discounted square sumable sequences of finite dimensional real vectors,

$$H = \left\{ (\zeta_t)_{t=\tau}^{\infty} \mid E \sum_{t=\tau}^{\infty} \beta^{t-\tau} \zeta_t' \zeta_t < \infty, \zeta_t \in R^n, n \in N_+ \right\}$$

Table 1: A priori restrictions imposed upon the structural parameters of the present model in selected studies

Study	Parameter								
	a	b	c	d	e	f	g	h	γ
Blanchard (1983)	*	+	+	*	0	+	+	∞	0
Blinder (1982)	0	+	0	0	+	0	0	+	0
Blinder (1986)	*	+	+	0	+	0	0	+	0
Eichenbaum (1983)	*	+	+	*	+	+	+	+	0
Eichenbaum (1984)	0	+	+	0	+	0	0	+	+
Holt et al. (1960)	0	+	+	0	+	+	+	∞	0
West (1986)	0	*	*	0	*	*	*	*	0

Key: *: no restrictions
      +: strictly positive
      0: zero
      ∞: infinity (i.e., cost minimization)

This property is crucial in ensuring that the supremum in (2.1) exists under appropriate restrictions on the growth of the endogenous variables.

Let $x_t = (i_t, q_{t-1})'$. The firm's information at the beginning of any period t, is the "history" of the $\{(x_t, \xi_t), t \in N_+\}$ process. That is, $\Omega_t = \{x_u, \xi_u\}_{u=0}^t$. A "contingency plan" $\{q_t, s_t\}_{t=\tau}^\infty$ specifies q_t and s_t as functions of the elements of Ω_t. That is, $q_t = q_t(\Omega_t)$ and $s_t = s_t(\Omega_t)$. With this structure of the stochastic environment and the firm's information, the expectations at any point in time t are with respect to the fixed distribution of the $\{\varepsilon_t : t \in N_+\}$ process $F(\varepsilon)$. We do not need to specify this distribution here because all we are going to do is to approximate these expectations by their linear minimum mean square error predictors.[8]

Finally, the expectations of firms are taken to be rational in the following sense. First, the firms subjective law of motion about the $\{\xi_t : t \in N_+\}$ and the objective law of motion of that process are identical (given by (2.7)); and second, the firms expectations about the $\{p_t, s_t\}_{t=\tau}^\infty$ process are such that the second requirement of the followig definition holds.

Definition: $\{p_t, i_{t+1}, q_t\}_{t=\tau}^\infty$ is said to be a rational expectations equilibrium if the following are true: (i) Given $\{p_t(\Omega_t)\}_{t=\tau}^\infty$, $q_t(\Omega_t)\}_{t=\tau}^\infty$ is a solution to the representative firm's problem, defined above; and (ii) Given $\{i_{t+1}(\Omega_t), q_t(\Omega_t)\}_{t=\tau}^\infty$, $\{p_t(\Omega_t)\}_{t=\tau}^\infty$ satisfies:

$$S_t = \frac{\eta_t}{h} - \frac{1}{h} p_t \text{ for } p_t \geq 0 \text{ and for all } t \in \{\tau, \tau+1, \ldots\}.$$

We proceed, now, to the solution of the equilibrium problem.[9]

3. The Solution of the Model

---

[8] Linear minimum mean square error predictors would have coincided with conditional expectations under some additional restrictions on the $\{\varepsilon_t : t \in N_+\}$ process. For example, if the $\{\varepsilon_t : t \in N_+\}$ is Gaussian. Here, we cannot assume that this process is Gaussian for the support of that process cannot be the entire plane R^2. For example, the support of the innovation of the aggregate demand shock cannot be the entire real line if prices are to be positive. This is the reason some authors (e.g., Christiano and Eichenbaum (1988)) take E to be the linear minimum mean square error prediction operator and $E_t(\cdot)$ to denote that predictions are linear functions of the elements of Ω_t.

[9] Note that we have chosen to represent the firm's plan by $\{i_{t+1}, q_t\}_{t=\tau}^\infty$ rather than $\{s_t, q_t\}_{t=\tau}^\infty$. Given (2.3) these two representations are equivalent. However, the former turns out to be more convenient.

3.1 Necessary Conditions

Let

$$V(i_\tau, q_{\tau-1}, \xi_\tau) = \sup_{\{i_{t+1}, q_t\}_{t=\tau}^\infty \in S(i_\tau, q_{\tau-1}, \xi_t)} E_\tau \sum_{t=\tau}^\infty \beta^{t-\tau} \Pi\{q_t - [i_{t+1}-(1-\gamma)i_t], q_t, i_t, q_{t-1}, \xi_t\} \quad (3.1)$$

where

$$S(i_\tau, q_{\tau-1}, \xi_\tau) = \left\{ \{i_{t+1}, q_t\}_{t=\tau}^\infty \mid \{i_{t+1}, q_t\}_{t=\tau}^\infty \in H, \; (q_t - [i_{t+1}-(1-\gamma)i_t], q_t) \geq 0, \; (x_\tau, \xi_\tau) \text{ given} \right\} \quad (3.2)$$

Note that the set of all feasible contingency plans, $S(i_\tau, q_{\tau-1}, \xi_\tau)$ is restricted in such a way that in view of the restrictions on the growth of $\{\xi_t : t \in N_+\}$, the value function $V(i_\tau, q_{\tau-1}, \xi_\tau)$ is well defined.[10]

Moreover, the growth restriction on $\{i_{t+1}, q_t\}_{t=\tau}^\infty$ implies that:

$$E_\tau \beta^{t-\tau} i_{t+1}, \; E_\tau \beta^{t-\tau} q_t \to 0 \text{ as } t \to \infty, \; \forall (i_\tau, q_{\tau-1}) \in R \times R_+ \quad (3.4)$$

This "global asymptotic stability" condition is stronger than the usual Transversality Condition.[11] But the cases where the Transversality Condition holds and (3.4) fails are essentially uninteresting. More on this in the next section.

---

[10] This is usually the most desirable way to ensure that the supremum in (3.1) exists. Typically, restrictions on β, Π and S are made that ensure that V is well defined in a compact set where (i_t, q_{t-1}, ξ_t) always lies if $\{i_{t+1}, q_t\}_{t=\tau}^\infty$ is to be optimal. (See, e.g., Stokey et al. (1987)). However, for our purposes this is not a strong assumption, for subsequently we will make restrictions that ensure that any optimal $\{i_{t+1}, q_t\}_{t=\tau}^\infty$ has this property. Further, if one allows $\{\xi_t, t \in N_+\}$ and $\{i_{t+1}, q_t\}_{t=\tau}^\infty$ to grow at faster rates, then he/she should, perhaps, employ an optimality criterion in the "overtaking" sense. See, e.g., Carlson and Haurie (1987).

[11] See (3.24) below.

Given our restriction on $\{i_{t+1}, s_t\}_{t=\tau}^{\infty}$ and $\{\xi_t : t \in \mathbb{N}_+\}$ it can be easily verified that $V(\cdot, \cdot, \cdot)$ satisfies the functional equation:

$$V(i_t, q_{t-1}, \xi_t) = \sup_{(i_{t+1}, q_t)} (\Pi\{q_t - [i_{t+1} - (1-\delta)i_t], q_t, i_t, q_{t-1}, \xi_t\}$$

$$+ \beta E_t V(i_{t+1}, q_t, \xi_{t+1})), \forall t \in \{\tau, \tau+1, \ldots\} \quad (3.5)$$

Moreover, we shall require that the following holds.[12]

<u>Assumption Ø</u>: There exists a function $V: \mathbb{R} \times \mathbb{R}_+ \times \mathbb{R} \to \mathbb{R}$ that satisfies (3.5) and moreover this function is twice differentiable in the interior of its domain.

This assumption would be typically too strong. However, in the present context it is not, for our subsequent solution will establish the existence, uniqueness, and differentiability of the value function. Then, if $\{i_{t+1}, q_t\}_{t=\tau}^{\infty}$ is an interior solution to (3.3), in the sense that the inequality constraints in $S(i_t, q_{t-1}, \xi_\tau)$ are not binding, we must have:

$$-\Pi_1 |^t + \beta E_t V_1 |^{t+1} = 0 \quad (3.6)$$

$$-\Pi_1 |^t + \Pi_2 |^t + \beta E_t V_2 |^{t+1} = 0 \quad (3.7)$$

$$\left[\Pi_{11} |^t + \beta E_t V_{11} |^{t+1} \qquad -\Pi_{11} |^t - \Pi_{12} |^t + \beta E_t V_{12} |^{t+1} \right]$$

---

[12]Alternatively, in the spirit of Footnote 10, we may proceed to formulate the equilibrium problem as the corresponding Social Planner's problem (See equation (3.19) below.) Then, we may impose a regularity condition ($|S| \neq 0$) to transform this problem in an Optimal Linear Regulator format. (See, e.g., Kwakernaak and Sivan (1972)). Then, we may make assumptions that ensure "detectability" and "stabilizability" to find the optimal solution and $V(\cdot, \cdot, \cdot)$ simultaneously. For example, the sufficiency condition (3.12) will ensure these conditions. This method is discussed in Kollintzas (1985, 1986b). This method ignores the nonnegativity restrictions. But, these nonnegativity restrictions can be ensured by means of restrictions on the innovation of the $\{\xi_t : t \in \mathbb{N}_+\}$ process.

$$\left[-\Pi_{11}|^t -\Pi_{21}|^t + \beta E_t V_{21}|^{t+1} \quad \Pi_{11}|^t + \Pi_{12}|^t + \Pi_{21}|^t + \Pi_{22}|^t + \beta E_t V_{22}|^{t+1} \right]$$

≥ 0 (i.e., negative semidefinite) \hfill (3.8)

where $\Pi_1|^t$ stands for the partial derivative of Π with respect to its first argument evaluated at $\{q_t - [i_{t+1} - (1-\gamma)i_t], q_t, i_t, q_{t-1}, \xi_t\}$, $V_1|^{t+1}$ stands for the partial derivative of V with respect to its first argument evaluated at $(i_{t+1}, q_t, \xi_{t+1})$, etc. Also, from (3.3) and by the Envelope Theorem, we have:

$$V_1|^{t+1} = (1-\gamma) \Pi_1|^{t+1} + \Pi_3|^{t+1} \tag{3.9}$$

$$V_2|^{t+1} = \Pi_4|^{t+1} \tag{3.10}$$

$$V_{11}|^{t+1} = (1-\gamma)^2 \Pi_{11}|^{t+1} + (1-\gamma) \Pi_{13}|^{t+1} + (1-\gamma) \Pi_{31}|^{t+1} + \Pi_{33}|^{t+1} \tag{3.11}$$

$$V_{12}|^{t+1} = (1-\gamma) \Pi_{14}|^{t+1} + \Pi_{34}|^{t+1} \tag{3.12}$$

$$V_{21}|^{t+1} = (1-\gamma) \Pi_{41}|^{t+1} + \Pi_{43}|^{t+1} \tag{3.13}$$

$$V_{22}|^{t+1} = \Pi_{44}|^{t+1} \tag{3.14}$$

Combining (3.4) - (3.12) implies that if $\{i_{t+1}, q_t\}_{t=\tau}^{\infty}$, is a solution to (3.3) we must have:

$$-[\Pi_2|^t + \beta E_t \Pi_4|^t] = \beta E_t [(1-\gamma) \Pi_1|^{t+1} + \Pi_3|^{t+1}] = \Pi_1|^t \tag{3.15}$$

and

$$\begin{bmatrix} \beta^{-1}\Pi_{11}|^t + (1-\gamma)\Pi_{11}|^{t+1} + (1-\gamma)\Pi_{13}|^{t+1}(1-\gamma)\Pi_{31}|^{t+1}\Pi_{33}|^t \\ -\beta^{-1}\Pi_{11}|^t - \beta^{-1}\Pi_{21}|^t + (1-\gamma)\Pi_{41}|^{t+1} + \Pi_{43}|^{t+1} \\ -\beta^{-1}\Pi_{11}|^t - \beta^{-1}\Pi_{12}|^t + (1-\gamma)\Pi_{14}|^{t+1} + \Pi_{34}|^{t+1} \\ \beta^{-1}\Pi_{11}|^t + \beta^{-1}\Pi_{12}|^t + \beta^{-1}\Pi_{22}|^t + \beta^{-1}\Pi_{22}|^t + \Pi_{44}|^{t+1} \end{bmatrix} \leq 0 \quad (3.16)$$

Equation (3.13) is an Euler Condition. This condition is a generalization of the celebrated result in Blinder (1982). According to this condition the firm should equate its marginal production cost in period t, $\Pi_1|^t$, adjusted for the discounted expected change in adjustment costs next period to the "expected shadow value of inventories" next period, $\beta E_t[(1-\delta)\Pi_2|^{t+1} + \Pi_3|^{t+1}]$ as well as its current marginal revenue, $\Pi_1|^t$. See Figure 1, below. $\Pi_2|^t$ has a positive slope as long as h is positive. That is, even if firms are competitive and take prices as given, they ought to realize that if they expand (contract) so does the industry and therefore they face a downward sloping demand schedule. Alternatively, (3.15) requires that the discounted future stream of net marginal benefits associated with producing and selling output in any given period as opposed doing so in the next period must be zero. Then, (3.16) which is a Legendre Condition, requires that the above mentioned benefits must be diminishing for any combination of additional production and sales.[13] The implications of the Legendre Condition for the parameters of the model will be made clear shortly.

Now, it follows that if $\{p_t, i_{t+1}, q_t\}_{t=\tau}^{\infty}$ is an "interior" rational expectations equilibrium in the sense that $(p_t, q_t - [i_{t+1} - (1-\gamma)i_t], q_t) \geq 0$, in addition to (3.15), (3.16), and (2.14), (2.6) must also hold. Note that by placing restrictions on the intercepts of the lines in Figure 1 these nonnegativity constraints can be ensured. As already noted this may be done by placing restrictions in the range of the innovations of the $\{\xi_t : t \in N_+\}$ process, $(\underline{\varepsilon}, \bar{\varepsilon})$. Then, in view of (2.2) and (2.6) we have the following:

---

[13] See, e.g., Kamien and Shwartz (1980).

Figure 1: Illustration of the Euler Condition when $\Pi_{33} = 0$.

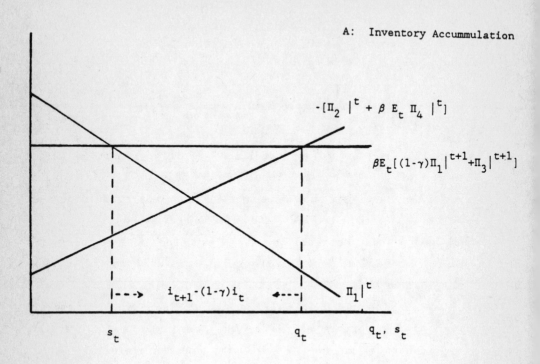

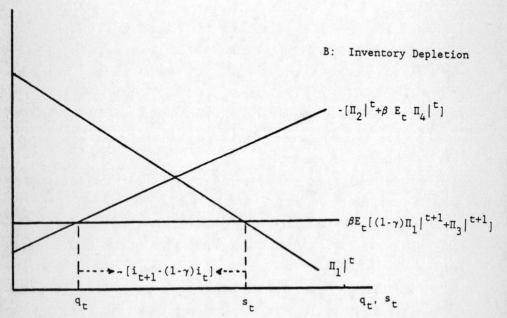

Proposition 1: If $\{p_t, i_{t+1}, q_t\}_{t=\tau}^{\infty}$ is an interior rational expectations equilibrium then the following conditions must hold:

$$\text{(S-R) } E_t x_{t+2} - [Q - (R+R') + (1+\beta^{-1})S] E_t x_{t+1} + \beta^{-1}(S-R')$$

$$E_t x_t = (O-P)' E_t \xi_{t+1} + \beta^{-1} P' E_t \xi_t \qquad (3.17)$$

$$Q - (R+R') + (1+\beta^{-1}) S \geq 0 \qquad (3.18)$$

where:

$$\Pi(s_t, q_t, i_t, q_{t-1}, \xi_t)\Big|_{p_t = \eta_t - 1/2\, Nhs_t} =$$

$$-1/2 \begin{bmatrix} \xi_t \\ x_t \\ x_{t+1}-x_t \end{bmatrix}' \begin{bmatrix} 0 & 0 & P \\ O' & Q & R \\ P' & R' & S \end{bmatrix} \begin{bmatrix} \xi_t \\ x_t \\ x_{t+1}-x_t \end{bmatrix} =$$

$$-1/2 \begin{bmatrix} \eta_t \\ \theta_t \\ i_t \\ q_{t-1} \\ i_{t+1}-i_t \\ q_t-q_{t-1} \end{bmatrix}' \begin{bmatrix} 0 & 0 & \gamma & -1 & 1 & -1 \\ 0 & d & a & 0 & a & \\ & & m & -(\gamma k+\ell) & \gamma k+\ell & -(\gamma k+\ell) \\ & & & k+b & -k & k+b \\ & & & & k & -k \\ & & & & & b+c+k \end{bmatrix} \begin{bmatrix} \eta_t \\ \theta_t \\ i_t \\ q_{t-1} \\ i_{t+1}-i_t \\ q_t-q_{t-1} \end{bmatrix} \qquad (3.19)$$

where

$$k = Nh + fg^2$$

$$\ell = fg$$

$$m = e + f + \gamma^2 k + 2\gamma \ell$$

Equation (3.17) is the Euler Condition and (3.18) is the Legendre Condition. Equation (3.19) is an implication of the well known result (Lucas and Prescott (1971)) that the necessary conditions for the equilibrium are those that correspond to the problem of a social planner that seeks to maximize the expected discounted future stream of consumer surplus minus the industry costs, or the problem of a monopolist that faces an aggregate (inverse) demand curve with half the slope of the aggregate (inverse) demand curve facing the competitive industry (Blinder (1982), Eichenbaum (1983)). Further, note that (3.18) requires

$$\begin{bmatrix} [\beta^{-1} + (1-\gamma)^2]k - 2(1-\gamma)\ell + e + f & -\beta^{-1}k \\ -\beta^{-1}k & \beta^{-1}(b+k) + (1+\beta^{-1})c \end{bmatrix} \geq 0 \quad (3.20)$$

Although (3.20) makes clear that some curvature restrictions must be made a priori, obviously, there are a number of cases where (3.20) holds and the strong curvature restrictions in Table 1 do not. For example, if profits do not increase at the margin with sales $k \geq 0$, (3.20) holds as long as inventory and backlog order costs are nondecreasing at the margin with inventory levels, $e + f \geq 0$ and as long as production and adjustment costs are nondecreasing at the margin. So with backlog and adjustment costs being sufficiently increasing at the margin, (3.20) allows for decreasing at the margin inventory and production costs. It is commonly believed that the cost minimization version of this model is a part of any profit maximization model irrespective of the aggregate demand and market structure. This is not true. And at this point it is appropriate to illustrate one of the differences between these two models that involves the Legendre Condition.

When $h \to \infty$ and $-\dfrac{\eta_t}{Nh} \to s_t = \eta_t$,

$$\Pi(s_t, q_t, i_t, q_{t-1}, \xi_t) = -1/2 \ \{$$

$$\begin{bmatrix} \eta_t \\ \theta_t \\ \eta_{t-1} \\ \theta_{t-1} \\ i_t \\ i_{t-1} \\ i_{t+1}-i_t \\ i_t-i_{t-1} \end{bmatrix}' \begin{bmatrix} b+c+fg^2 & a & -c & 0 & \gamma(b+c-fg) & -\gamma c & b+c & -c \\ & 0 & 0 & 0 & \gamma a+d & 0 & a & 0 \\ & & c & 0 & -\gamma c & \gamma c & -c & c \\ & & & 0 & 0 & 0 & 0 & 0 \\ & & & & \gamma^2 b+c+e+f & -\gamma^2 c & \gamma(b+c) & -\gamma c \\ & & & & & \gamma^2 c & -\gamma c & \gamma c \\ & & & & & & b+c & -c \\ & & & & & & & c^2 \end{bmatrix}$$

$$\begin{bmatrix} \eta_t \\ \theta_t \\ \eta_{t-1} \\ \theta_{t-1} \\ i_t \\ i_{t-1} \\ i_{t+1}-i_t \\ i_t-i_{t-1} \end{bmatrix} \quad \} \qquad (3.19)$$

Thus,

$$Q = \begin{bmatrix} \gamma^2(b+c)+e+f & -\gamma^2 c \\ -\gamma^2 c & \gamma^2 c \end{bmatrix}$$

$$R = \begin{bmatrix} \gamma(b+c) & -\gamma c \\ -\gamma c & \gamma c \end{bmatrix}$$

and

$$S = \begin{bmatrix} b+c & c \\ -c & c^2 \end{bmatrix}$$

The Legendre Condition for this problem is:

$$Q-R-R'+(1+\beta^{-1})\,S = \begin{bmatrix} [(1-\gamma)^2+\beta^{-1}](b+c)+e+f & -[(1-\gamma)^2+\beta^{-1}]c \\ -[(1-\gamma)^2+\beta^{-1}]c & [(1-\gamma)^2+\beta^{-1}]c \end{bmatrix} \geq 0 \quad (3.22)$$

Unlike condition (3.20), condition (3.22) requires adjustment costs to be nondecreasing at the margin, $c \geq 0$, and

$$e+f + [(1-\gamma)^2 + \beta^{-1}]b \geq 0$$

The last restriction implies that the expected discounted future stream of cost savings associated with producing this period rather than next period should be decreasing at the margin. Although this condition is also different from anything in (3.20), it also allows for decreasing at the margin production ($b < 0$) and inventory holding costs ($e < 0$).[14] More substantial differences between the profit and cost minimization models follow.

3.2 Sufficient Conditions

It follows by a standard argument that:[15]

Proposition 2: If $\{p_t, i_{t+1}, q_t\}_{t=\tau}^{\infty}$ is feasible, in the sense that $\{i_{t+1}, q_t\}_{t=\tau}^{\infty} \in S\,(i_\tau, q_{\tau-1}, \xi_\tau)$ and $p_t \geq 0$ satisfies (3.17), and

$$\begin{bmatrix} Q & R \\ R' & S \end{bmatrix} \geq 0 \quad (3.23)$$

---

[14] As discussed in Blinder (1986) and shown in Ramey (1987), decreasing at the margin production costs give rise to the possibility that production may vary more than sales, even when technology shocks and backlog costs are ignored. More on this later.

[15] See, e.g., the Appendix in Kollintzas (1988).

then it is an interior rational expectations equilibrium.

The difference between Proposition 2 and the usual sufficiency conditions is the absence of the Transversality Condition. The underlying Transversality Condition for this problem is:

$$E_\tau \beta^{t-\tau} \{S \, x(t+1)(S - R') x(t) + P'\xi(t)\}' \delta(t+1) - 1/2 \, \delta(t+1)'$$

$$S\delta(t+1)\} \to 0 \text{ as } t \to \infty, \; \forall \; \{\delta(t)\}_{t=\tau}^{\infty} \in H \text{ such that } \delta(\tau) = 0 \quad (3.24)^{16}$$

But since $\{x_t\}_{t=\tau}^{\infty}, \{\xi_t\}_{t=\tau}^{\infty} \in H$ it follows by the Cauchy-Schwartz inequality that (3.24) is automatically satisfied.[17] In the next subsection we shall analyze the restrictions that are necessary and sufficient to ensure that the particular $\{x_t\}_{t=\tau}^{\infty}$ which solves (3.17) is a element of H. First, however, we shall analyze (3.23).

Since

$$S = \begin{bmatrix} k & -k \\ -k & b+k+c \end{bmatrix}, \quad |S| = k(b+c).$$

So that if $k, (b+c) \neq 0$, $|S| \neq 0$ and

$$\begin{bmatrix} Q & R \\ R' & S \end{bmatrix} \geq 0 \qquad \text{if and only if}$$

$$\begin{bmatrix} I & 0 \\ -S^{-1}R' & I \end{bmatrix}' \begin{bmatrix} Q & R \\ R' & S \end{bmatrix} \begin{bmatrix} I & 0 \\ -S^{-1}R' & I \end{bmatrix} = \begin{bmatrix} Q - RS^{-1}R' & 0 \\ 0 & S \end{bmatrix} \geq 0.$$

Hence,

$$\begin{bmatrix} Q & R \\ R' & S \end{bmatrix} \geq 0 \qquad \text{if and only if}$$

---

[16] If we require $x(t+1) \geq 0$, $\forall \; t \geq \tau$, $x(t+1)$ may be substituted for $\delta(t+1)$ in (3.24). Compare this with the Transversality Conditions in Eichenbaum (1983, 1984) and Ramey (1987).

[17] It can be easily verified that H is a Hilbert space.

$$Q - RS^{-1}R' = \begin{bmatrix} e+f-\ell^2/k & 0 \\ 0 & bc/(b+c) \end{bmatrix} \geq 0 \text{ and } S > 0.$$

Summarizing results, we have the following.

<u>Corollary</u>: Given k, $(b+c) \neq 0$

$$\begin{bmatrix} Q & R \\ R' & S \end{bmatrix} \geq 0 \quad \text{if and only if}$$

$(b+c)$, $k > 0$ and b, c, $e+f-\ell^2/k \geq 0$

Given k, $(b+c) \neq 0$, sufficiency requires: (i) The rate at which production and adjustment costs increase jointly with production to be positive and the rates at which these costs increase individually with production to be nonnegative; (ii) The rate at which profits increase with sales to be negative; and (iii) The rate at which inventory holding costs increase with inventories to be no less than a possibly negative ratio bound $\ell^2/k - f$. Thus, the curvature restrictions imposed by most authors in the literature will suffice for sufficiency.[18] It is remarkable how much stronger these conditions are relative to the conditions of the previous subsection.

3.3 General Solution

The general solution to the equilibrium problem may be obtained in two steps. First, one may obtain the general solution to (3.17) and then he/she may obtain the particular form of that solution so as to satisfy feasibility. In order to obtain the general solution to (3.17), first, the characteristic polynomial associated with (3.17):

$$\phi(\lambda) = (S-R)\lambda^2 - [Q-(R+R')+(1+\beta^{-1})S]\lambda + \beta^{-1}(S-R)' \quad (3.25)$$

must be factored. Since

---

[18] An alternative set of conditions can be derived when $Q > 0$. We leave this case as well as the sufficient conditions in the cost minimization case an an exercise for the reader.

$$S-R = \begin{bmatrix} (1-\gamma)k-\ell & -[(1-\gamma)k-\ell] \\ 0 & c \end{bmatrix}$$

is not symmetric there is not, in general, any simple method for factoring out $\phi(\lambda)$.[19] For our purposes, a suitable method for factoring out $\phi(\lambda)$ is the specialization of the Blanchard and Kahn (1981) method developed in Kollintzas (1986a, 1986b). Heuristically, this method consists of finding the eigenvalues and the eigenvectors of $\phi(\lambda)$. Then, under a mild restriction on the nature of the eigenvalues and eigenvectors of $\phi(\lambda)$ it is possible to define two right divisors of $\phi(\lambda)$ with each one of these divisions being associated with a matrix that its eigenvalues are in modulus less than or equal to $\beta^{-1/2}$ and a matrix that its eigenvalues are in modulus greater than or equal to $\beta^{-1/2}$. Moreover, these matrices have no common eigenvalues and this implies that their difference is nonsingular. It turns out that this property allows for a convenient factorization of $\phi(\lambda)$ and hence gives us an easy way to obtain the general solution to (3.17). To apply this method we need the following assumptions.

<u>Assumption 1</u>: $c, [(1-\gamma)k-\ell] \neq 0$

This assumption implies that $|S-R| \neq 0$. The implication of this is that if $\lambda \in \mathcal{C}$ is a eigenvalue of λ then so is $(\beta\lambda)^{-1} \in \mathcal{C}$. This assumption is not really crucial because if c or $[(1-\gamma)k-\ell] = 0$ then the equilibrium problem is of lower dimension and the rest of the analysis may proceed by solving an appropriate univariate version of (3.17) as in Blinder (1982, 1986). Now, we have the following:

<u>Lemma 1</u>: Let λ_1, λ_2, λ_3, λ_4 denote the eigenvalues of $\phi(\lambda)$. Then, given Assumption 1, one may take:

---

[19] In particular one cannot employ the method of Kollintzas (1985).

$$\lambda_1 \cdot \lambda_2 = \beta^{-1} \quad \& \quad \lambda_1 + \lambda_2 = -\frac{\sigma}{2} + \sqrt{(\frac{\sigma}{2})^2 - \rho}$$

$$\lambda_3 \cdot \lambda_4 = \beta^{-1} \quad \& \quad \lambda_3 + \lambda_4 = -\frac{\sigma}{2} - \sqrt{(\frac{\sigma}{2})^2 - \rho}$$

$$\sigma = \frac{\{e+f+[(1-\gamma)^2+\beta^{-1}]k-2(1-\gamma)\ell\}c+[(1+\beta^{-1})c+\beta^{-1}b][(1-\gamma)k-\ell]}{c\,[(1-\gamma)k-\ell]}$$

$$\rho = \frac{\{e+f+[(1-\gamma)^2+\beta^{-1}]k-2(1-\gamma)\ell\}[(1+\beta^{-1})c+\beta^{-1}(b+k)] - (\beta^{-1}k)^2}{c\,[(1-\gamma)k-\ell]}$$

The usefulness of this result is that one can characterize the nature of the λ_i's based on β, σ, and ρ alone.

$$\mathcal{R}_{1A} = \{(-\tfrac{\sigma}{2}, \rho) \in R^2 | \rho > (-\tfrac{\sigma}{2})^2\}$$

$$\mathcal{R}_{1B} = \{(-\tfrac{\sigma}{2}, \rho) \in R^2 | \rho \leq (-\tfrac{\sigma}{2})^2 \ \& \ \rho > -4\beta^{-1} \pm 4\beta^{-1/2}(-\tfrac{\sigma}{2})\}$$

$$\mathcal{R}_{2A} = \{(-\tfrac{\sigma}{2}, \rho) \in R^2 | \rho \in (4\beta^{-1}, \infty) \ \& \ \rho = (-\tfrac{\sigma}{2})^2\}$$

$$\mathcal{R}_{2B} = \{(-\tfrac{\sigma}{2}, \rho) \in R^2 | \rho \in [4\beta^{-1}, \infty) \cup (-\infty, -4\beta^{-1}) \ \&$$

$$\rho = -4\beta^{-1} \pm 4\beta^{-1/2}(-\tfrac{\sigma}{2})\}$$

$$\mathcal{R}_{2C} = \{(-\tfrac{\sigma}{2}, \rho) \in R^2 | [-4\beta^{-1}, 4\beta^{-1}) \ \& \ \rho = -4\beta^{-1} \pm 4\beta^{-1/2}(-\tfrac{\sigma}{2})\}$$

$$\mathcal{R}_{3A} = \{(-\tfrac{\sigma}{2}, \rho) \in R^2 | \rho < -4\beta^{-1} + 4\beta^{-1/2}(-\tfrac{\sigma}{2})\}\backslash R_{4C}$$

$$\mathcal{R}_{3B} = \{(-\tfrac{\sigma}{2}, \rho) \in R^2 | \rho < -4\beta^{-1} + 4\beta^{-1/2}(-\tfrac{\sigma}{2})\}\backslash R_{4C}$$

$$\mathcal{R}_{4A} = \{(-\tfrac{\sigma}{2}, \rho) \in R^2 | \rho \in (4\beta^{-1}, \infty) \ (-\tfrac{\sigma}{2})^2 > \rho > -4\beta^{-1}+4\beta^{-1/2}(-\tfrac{\sigma}{2})\}$$

$$\Re_{4B} = \{(-\tfrac{\sigma}{2}, \rho) \in R^2 | \rho \in (4\beta^{-1},\infty) \ \& \ -4\beta^{-1}-4\beta^{-1/2}(-\tfrac{\sigma}{2}) < \rho < (-\tfrac{\sigma}{2})^2\}$$

$$\Re_{4C} \equiv \{(-\tfrac{\sigma}{2}, \rho) \in R^2 | \rho \in (-\infty, -4\beta^{-1}) \ \& \ \rho < -4\beta^{-1} \pm 4\beta^{-1/2}(-\tfrac{\sigma}{2})\}$$

Note that these regions form a partition of R^2. See Figure 2 below. Then, the nature of the eigenvalues of $\phi(\lambda)$ is completely characterized by the following:

<u>Lemma 2</u>: Let $\lambda_1, \lambda_2, \lambda_3, \lambda_4$ be as in Lemma 1. Then,

$\lambda_1 = \bar{\lambda}_3 \ \& \ \lambda_2 = \bar{\lambda}_4$ when $(-\tfrac{\sigma}{2}, \rho) \in \Re_{1A}$

$\lambda_1 = \bar{\lambda}_2 \ \& \ \lambda_3 = \bar{\lambda}_4$ when $(-\tfrac{\sigma}{2}, \rho) \in \Re_{1B}$

$\lambda_1 = \bar{\lambda}_3 \in \Re \ \& \ \lambda_2 = \bar{\lambda}_4 \in \Re$ when $(-\tfrac{\sigma}{2}, \rho) \in \Re_{2A}$

$\lambda_1, \lambda_2, \lambda_3, \lambda_4 \in \Re \ \& \ \lambda_1 = \lambda_2$ or $\lambda_3 = \lambda_4$ when $(-\tfrac{\sigma}{2}, \rho) \in \Re_{2B}$

$\lambda_1 = \bar{\lambda}_2 \ \& \ \lambda_3 = \lambda_4 \in \Re$ or $\lambda_3 = \bar{\lambda}_4 \ \& \ \lambda_1 = \lambda_2 \in \Re$ when $(-\tfrac{\sigma}{2}, \rho) \in \Re_{2C}$

$\lambda_1 = \bar{\lambda}_2 \ \& \ \lambda_3 \neq \lambda_4 \in R$ when $(-\tfrac{\sigma}{2}, \rho) \in \Re_{3A}$

$\lambda_3 = \bar{\lambda}_4 \ \& \ \lambda_1 \neq \lambda_2 \in \Re$ when $(-\tfrac{\sigma}{2}, \rho) \in \Re_{3B}$

$\lambda_1, \lambda_2, \lambda_3, \lambda_4 \in R \ \& \ \lambda_i \neq \lambda_j, \forall_{i,j} = 1,2,3,4$ when $(-\tfrac{\sigma}{2}, \rho) \in \Re_4 \equiv$

$\Re_{4A} \cup \Re_{4B} \cup \Re_{4C}$.

Again, for convenience, we rule out the case of repeated eigenvalues.

<u>Assumption 2</u>: $\rho \neq (-\tfrac{\sigma}{2})^2 \ \& \ \rho \neq -4\beta^{-1} \pm 4\beta^{-1/2}(-\tfrac{\sigma}{2})$

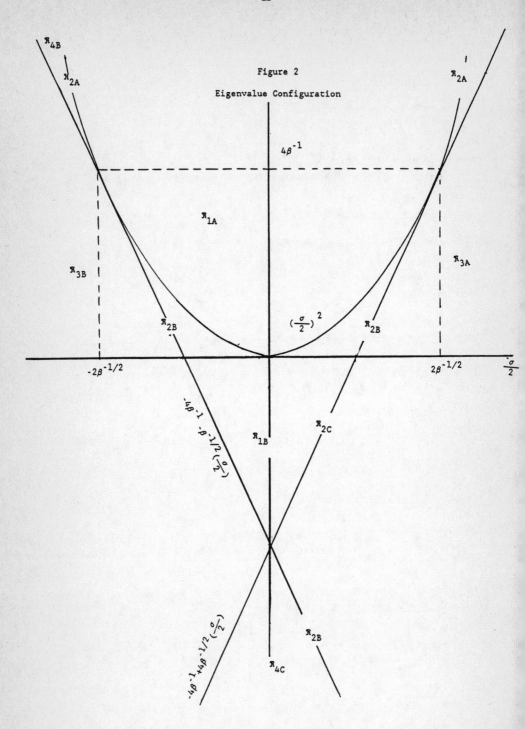

Figure 2
Eigenvalue Configuration

This is made for simplicity. The results of this paper can be extended to the case of repeated eigenvalues as in Kollintzas (1986a, 1986b).[20] But, now, one can split the eigenvalues of λ in those with modulus less than or equal to $\beta^{-1/2}$, say, λ_1 and λ_2, and those with modulus equal or greater than $\beta^{-1/2}$, $\beta^{-1}\lambda_1^{-1}$ and $\beta^{-1}\lambda_2^{-1}$. Now consider the matrix with the eigenvectors of $\phi(\lambda)$ associated with $\Lambda = \text{diag } \{\lambda_1, \lambda_2\}$,

$$M = \begin{bmatrix} \mu_{11} & \mu_{12} \\ \mu_{21} & \mu_{22} \end{bmatrix} \text{ and}$$

the matrix with the eigenvectors associated with $\beta^{-1}\Lambda^{-1} = \text{diag } \{\beta^{-1}\lambda_1^{-1}, \beta^{-1}\lambda_2^{-1}\}$,

$$N = \begin{bmatrix} \nu_{11} & \nu_{12} \\ \nu_{21} & \nu_{22} \end{bmatrix}$$

By construction,

$$(S-R)M \Lambda^2 - [Q-(R+R') + (1+\beta^{-1})S] M \Lambda + \beta^{-1}(S-R)' M = 0 \quad (3.26)$$

$$(S-R)N(\beta^{-1}\Lambda^{-1})^2 - [Q-(R+R')+(1+\beta^{-1})S] N(\beta^{-1}\Lambda^{-1})+\beta^{-1}(S-R)'N = 0 \quad (3.27)$$

Matrices M and N are not, in general, nonsingular. However, given Assumptions 1 and 2, M and N turn out to be nonsingular.

Lemma 3: Given Assumptions 1 and 2, M and N, defined above, are nonsingular.

Proof: Suppose that M is singular. Then, $\mu_1 = (\mu_{11}, \mu_{21})'$ and $\mu_2 = (\mu_{21}, \mu_{22})$ are linearly dependent. Then, there exists an $\alpha \in R\backslash\{0\}$ such that $\mu_2 = \alpha\mu_1$. By construction, $[(S-R)\lambda_1^2 - [Q - (R + R') + (1+ \beta^{-1})$

---

[20] To avoid repetition, only the results that are not available in Kollintzas (1986a, 1986b) are proved here.

$S]\lambda_1 + \beta^{-1}(S - R')]\mu_1 = [(S-R)\lambda_2^2 - [Q - (R+R') + (1+\beta^{-1})S]\lambda_2 + \beta^{-1}(S-R')] \alpha \mu_1 = 0$. Then, $[Q - (R + R') + (1 + \beta^{-1})S]\mu_1 = [(S - R)\lambda_1 + \beta^{-1}\lambda_1^{-1}(S - R')]\mu_1 = [(S - R)\lambda_2 + \beta^{-1}\lambda_2^{-1}(S - R')]\mu_1$, since α, and by Assumption 1, $\lambda_1, \lambda_2 \neq 0$. But the last equality implies $(S - R)(\lambda_2 - \lambda_1) = (S - R')\beta\lambda_1^{-1}\lambda_2^{-1}(\lambda_2 - \lambda_1)$ which, in view of Assumption 2, implies

$$S - R = \begin{bmatrix} (1-\gamma)k-\ell & -[(1-\gamma)k-\ell] \\ 0 & c \end{bmatrix} =$$

$$\beta\lambda_1^{-1}\lambda_2^{-1} \begin{bmatrix} [(1-\gamma)k-\ell] & 0 \\ -[(1-\gamma)k-\ell] & c \end{bmatrix} = (S - R')\beta\lambda_1^{-1}\lambda_2$$

Which is, of course, impossible. Hence M cannot be nonsingular. Likewise for N.

Q.E.D.

Completing our linear rational expectations set up we make the following:

<u>Assumption 3</u>: $\beta^{j/2} E_t \xi(t+j) = (\beta^{1/2}A)^j \xi(t), \forall_j \in N_+$

That is, weighted conditional expectations are formally replaced by minimum mean square error predictors. With this result and assumption we may state the following:

<u>Lemma 4</u>: Given Assumptions 1 - 3, $|N\beta^{-1}\wedge^{-1}N^{-1}-MNM^{-1}| \neq 0$ and the solution to (3.17) is given by:

$$x_{t+1} = T x_t + U \xi_t + V^{-t} \hat{x} \qquad (3.28)$$

where:

$$T = M \wedge M^{-1}$$

$$V = (N\beta^{-1}\wedge^{-1}N^{-1} - M\wedge M^{-1}) N \beta \wedge N^{-1}(N\beta^{-1}\wedge^{-1}N^{-1} - M\wedge M^{-1})^{-1}$$

$$U = V(S-R)^{-1}\beta^{-1}P' + \sum_{j=1}^{\infty} V^j [(S-R)^{-1} (O-P)' + V(S-R)^{-1} \beta^{-1}P']A^j$$

and $\hat{x} \in \phi^2$ is arbitrary.

3.4 Globally Asympotically Stable Solution

It remains to find the particular solution to (3.17) that satisfies $\{x_t\}_{t=\tau}^{\infty} \in H$. Clearly then $\{x_t\}_{t=\tau}^{\infty}$ must satisfy (3.28) and (3.4). But as it turns out, if this is true then $\{x_t\}_{t=\tau}^{\infty} \in H$. Thus our task simplifies to find under what conditions (3.28) is globally assymptotically stable (i.e., satisfies (3.4)).

The following is straightforward.

Proposition 3: (3.28) satisfies (3.4) if and only if the eigenvalues of T are in modulus less than $\beta^{-1/2}$ and $\hat{x} = 0$.

For this reason it is required that this holds:

Assumption 4: The eigenvalues of T are in modulus less than $\beta^{-1/2}$.

Obviously, if the eigenvalues of T are in modulus less than $\beta^{-1/2}$ then so must be the eigenvalues of V. Now, we may state the solution to the equilibrium.

Proposition 4: Given Assumptions 1-4 the unique interior rational expectations equilibrium is given by (2.3)-(2.6) and

$$x_{t+1} = T x_t + U\xi_t \qquad (3.29)[21]$$

4. Stability, Cyclical Motion, and Comparative Dynamics.

---

[21] As promised, from this result it follows that $V(x_t,\xi_t)$ is a well defined quadratic in (x_t,ξ_t) and hence twice differentiable.

The preceding results and the following two lemmas completely characterize the properties of the eigenvalues of $\phi(\Lambda)$ permitted in view of Assumption 4.[22]

Lemma 4: Given Assumptions 1 and 2, Assumption 4 holds if and only if:

$$(-\tfrac{\sigma}{2}, \rho) \in \mathcal{R}_{1A} \cup \mathcal{R}_{2A} \cup \mathcal{R}_4$$

where $\mathcal{R}_4 = \mathcal{R}_{4A} \cup \mathcal{R}_{4B} \cup \mathcal{R}_{4C}$

Lemma 5

$$\text{sign Re}(\lambda_1) = \text{sign Re}(\lambda_2) = \begin{cases} - &, (-\tfrac{\sigma}{2},\rho) \in R_- \times R_+ \\ + &, \text{otherwise} \end{cases}$$

and

$$\text{sign Re}(\lambda_3) = \text{sign Re}(\lambda_1) = \begin{cases} + &, (-\tfrac{\sigma}{2},\rho) \in R_+ \times R_+ \\ - &, \text{otherwise} \end{cases}$$

where $R_- = (-\infty, 0]$.

In this section we make use of these results to characterize the stability, cyclical motion, and comparative dynamics properties of the equilibrium.

We shall say that the equilibrium exhibits periodic motion, cyclical motion, mixed motion, or monotone motion when λ_1, $\lambda_2 \in R_-$ λ_1, $\lambda_2 \in \mathcal{C}$, λ_1, $\lambda_2 \in R$ & $\lambda_1 \cdot \lambda_2 < 0$, λ_1, $\lambda_2 \in R_+$, respectively.

Then it follows from Lemmas 4 and 5 that the following is true:

Proposition 5: If

---

[22] Clearly Assumption 4 implies Assumptions 1 and 2.

$$(\tfrac{\sigma}{2},\rho)\epsilon \begin{cases} \mathfrak{R}_{1A}, & \text{system exhibits cyclical motion} \\ \mathfrak{R}_{4b}, & \text{system exhibits periodical motion} \\ \mathfrak{R}_{4c}, & \text{system exhibits mixed motion} \\ \mathfrak{R}_{4A}, & \text{system exhibits monotone motion} \end{cases}$$

It is important and somewhat surprising to realize that the cost minimization model does not give rise to stable cyclical motion.[23] To see this, note that in this case $R' = R$ and it follows from the result in Kollintzas (1985) that Assumption 4 implies $\lambda_1, \lambda_2 \in R$. The easiest case to see this is when $h = \infty$, $c = 0$, and $\gamma = 0$, and therefore:

$$\phi(\lambda) = b\lambda^2 - [(1+\beta^{-1})b + \ell+f]\lambda + \beta^{-1}b$$

and clearly, $|\lambda| < \beta^{-1/2}$ if and only if $\lambda_1, \lambda_2 \in R$ or if and only if

$$(e+f) > \begin{cases} -(\beta^{-1/2}-1)b, & b > 0 \\ (\beta^{-1/2}+1)b, & b < 0 \end{cases}$$

That $b > 0$ is not required illustrates that the restrictions imposed by Assumption 4 are not as strong as the sufficiency conditions that are usually invoked in the literature. The interpretation of Assumption 4 is, in general, that it imposes bounds on the rates at which stocks changes can be brought forward or postponed. See Kollintzas (1985 and 1986b).

Now, the preceding results on deterministic motion become easier to understand if we give the equilibrium the flexible accelerator representation:

<u>Corollary 2 (Lovel (1961))</u>: If $|T - I| \neq 0$,

---

[23] Compare this with the corresponding results in Blanchard (1983) and Ramey (1987). It seems however that if dynamic interrelations between adjustment, production or inventory holding costs are introduced as in Eichenbaum (1983, 1984) this result may change.

$\mathfrak{R}$

$$x(t+1) - x(t) = (I - T) [x^*(t) - x(t)]$$

where

$$x^*(t) = (I - T)^{-1} U \xi(t).$$

$x^*(t) = (i(t)^*, q(t-1)^*)'$ may be thought of as the vector with the levels of "desired" inventories and production. These results provide new information to the production smoothing issue by making it a non-issue. That is, as long as the eigenvalues of T are in modulus greater than 1 (i.e., between 1 and $\beta^{-1/2}$), or if they are negative, or if they are complex, it does not follow that production would adjust smoothhly to a change in desired inventories and production. Periodic and/or cyclical fluctuations are possible.

Under the sufficient conditions of Proposition 2 it is only possible to have eigenvalues that are real, positive, and less than 1. But, under the Legendre Condition and the Stability Condition (Assumption 4), all other non-smooth behavior possibilities exist. As evidenced from the cost minimization example, above, if firms have declining marginal production costs output and inventories may fluctuate. However, deterministic fluctuations can also occur when marginal production costs are rising.[24] That is, in case $\mathfrak{R}_{1A}$. This may seem paradoxical at first. Why can't firms lower costs by shifting production over to periods with relatively low marginal costs? The answer seems to be, that even if firms could lower costs by shifting production over to these periods, they would not increase profits. If inventories facilitate sales, in the sense of increasing marginal revenues ($\eta_t - ks_t + li_t$), firms have an incentive to enter periods of high sales with high inventories and vice versa. And, periods with high sales may come about when the increase in production and adjustment costs next period are less than the current increase in revenues.

---

[24] For example, when $\beta = .95$, $\gamma = 0$, $b = .5$, $c = .5$, $e = -.1$, $k = .2$, $1 = .01$, $f = .3$, $\lambda_1, \lambda_2 = 0.3325 \pm 0.1250$ i and $\lambda_3, \lambda_4 = 2.7742 \pm 1.0426$ i.

Thereby giving rise to a pattern of production, sales, and inventories as follows:

	period t	period t+1	period t+2
production	low	high	low
sales	high	low	high
inventories	high	low	high

These and other issues pertaining to stability, cyclical motion and comparative dynamics should be investigated further. Two things seem clear, however. The view that the linear rational expectations equilibrium inventory model is in trouble because it cannot account for the variance of production being greater than the variance of sales and inventories being procyclical, in the absence of cost shocks, is misleading. And, suppressing cyclical motion by a priori restrictions, may seriously overestimate the role of stochastic shocks.

5. Concluding Remarks

The linear rational expectations equilibrium inventory model is a modification of the model of Holt et al. (1960). This model forms the basis for many business cycle studies as well as industrial organization studies. However, despite its popularity, this model is not fully understood and sometimes mishandled. Primary responsibility for this anomaly lies with the preoccupation of studying and testing the Euler Condition for its solution and ignoring the other necessary conditions (Legendre and Transversality/Stability) and their implications. Further, the deterministic dynamic properties of the model have been ignored except in the simplest cases. The possibility of endogenous cycling and periodic motion have been ignored, despite their obvious importance for business cycle theory and stabilization policy.

This paper develops the necessary and sufficient conditions for a solution to the linear rational expectations equilibrium model and proceeds to solve that model and study the deterministic dynamic properties of that solution. Curvature necessary conditions are the Legendre and the Stability Conditions. Both of these conditions do not rule out diminishing marginal production, adjustment, or inventory holdig costs. Curvature sufficient conditions rule out all of the above except diminishing inventory holding costs in the presence of rising backlog costs. Most importantly, when these sufficient conditions hold, production and inventory behavior is smoothing. However, under the necessary conditions, cyclical as well as periodic behavior is possible.

Further analysis of the issues raised here is warranted. Especially, future studies should examine the role of production lags, stockouts, other factors of production.[25] Last but not least, more structure based on optimizing behavior and rational expectations should be imposed on the demand side of the model.

---

[25] Production lags should be similar to our adjustment costs setup. Stockouts have been considered in Aiyagari et al. (1988), Abel (1985), and Kahn (1987). Factors of production have been considered explicitly by several authors and most notably by Maccini (1984).

References

Abel, Andrew B. (1985): "Inventories, Stock-Outs and Production Smoothing," Review of Economic Studies, 52, 283-293.

Aiyagari, Rao, Zvi Eckstein, and Martin Eichenbaum (1988): "Inventories and Price Fluctuations Under Perfect Competition and Monopoly," Chapter II of this book.

Blanchard, Olivier J. (1983): "The Production and Inventory Behavior of the American Automobile Industry," Journal of Political Economy, 31, 365-400.

Blanchard, Olivier G. and Charles M. Kahn (1980): The Solution of Linear Difference Models Under Rational Expectations, Econometrica, 48, 1305-13011.

Blinder, Alan S. (1982): "Inventories and Sticky Prices: More On the Microfoundations of Macroeconomics," American Economic Review, 72,334-348.

Blinder, Alan S. (1986): "Can the Production Smoothing Model of Inventory be Saved?", Quarterly Journal of Economics, 101,431-454.

Carlson, D. A. and A. Haurie (1987): Infinite Horizon Optimal Control: Theory and Applications, Berlin: Springer-Vezlag.

Christiano, Lawrence T. and Martin Eichenbaum (1988): "Temporal Aggregation and the Stock Adjustment Model of Inventories," Chapter III of this book.

Eichenbaum, Martin (1983): "A Rational Expectations Equilibrium Model of Inventories of Finished Goods and Employment," Journal of Monetary Economics, 12, 259-277.

Eichenbaum, Martin (1984): "Rational Expectations and the Smoothing Properties of Inventories of Finished Goods," Journal of Monetary Economics, 14, 71-96.

Eichenbaum, Martin (1988): "Some Empirial Evidence on the Production Level and Production Cost Smoothing Models of Inventory Investment," Working Paper No. 2523, National Bureau of Economic Research.

Hansen, Lars P., and Thomas J. Sargent (1980): "Formulating and Estimating Dynamic Linear Rational Expectations Models," Journal of Dynamic Economics and Control, 2, 7-46.

Holt, Charles C., Franco Modigliani, John F. Muth, and Herbert A. Simon (1960): Planning Production, Inventories, and Work Force, Englewood Cliffs, N.J.: Prentice Hall.

Kahn, James A. (1987): "Inventories and the Volatility of Production," American Economic Review, 77, 667-679.

Kamien, Morton I. and Nancy L. Schwartz (1981): <u>Dynamic Optimization: The Calculus of Variables and Optimal Control in Economics and Management</u>, New York: North Holland.

Kollintzas, Tryphon (1985): "The Symmetric Linear Rational Expectations Model," <u>Econometrica</u>, 53, 963-976.

Kollintzas, Tryphon (1986a): "A Nonrecursive Solution for the Linear Rational Expectations Model," <u>Journal of Economic Dynamics and Control</u>, 10, 327-332.

Kollintzas, Tryphon (1986b): "Some New Results Concerning the Solution to Linear Rational Expectations Models Arising from Optimizing Behavior," Working Paper No. 204, Department of Economics, University of Pittsburgh.

Kollintzas, Tryphon (1988): "A Generalized Variance Bounds Test," Staff Report 113, Research Department, Federal Reserve Bank of Minneapolis.

Kollintzas, Tryphon and Harrie L.A.M. Geerts (1984): "Three Notes on the Formulation of Linear Rational Expectations Models," in <u>Time Series Analysis: Theory and Practice 5</u>, ed. by O. D. Anderson, Amsterdam: North Holland.

Kwakernaak, Huibert and Raphael Sivan (1972): <u>Linear Optimal Control Systems</u>, New York: John Wiley.

Lovell, Michael (1961): "Manufacturers' Inventories Sales Expectations, and the Accelerator Principle," <u>Econometrica</u>, 29, 293-314.

Lucas, Robert E. and Edward C. Prescott: "Investment under Uncertainty," <u>Econometrica</u>, 39, 659-668.

Maccini, Louis J. (1984): "The Interrelationship Between Price and Output Decisions and Investment Decisions," <u>Journal of Monetary Economics</u>, 13, 41-65.

Ramey, Valerie A. (1987): "Non-Convex Costs and the Behavior of Inventories'" Manuscript, University of California, San Diego.

Stokey, Nancy L., Robert E. Lucas, and Edward C. Prescott (1987): "Recursive Methods in Economic Dynamics," Manuscript, Northwestern University.

West, Kenneth D. (1986): "A Variance Bounds Test of the Linear Quadratic Inventory Model," <u>Journal of Political Economy</u>, 94, 374-401.

West, Kenneth D. (1987): "The Sources of Fluctuations in Aggregate GNP and Inventories," Manuscript, Princeton University.

CHAPTER II

INVENTORIES AND PRICE FLUCTUATIONS UNDER PERFECT COMPETITION AND MONOPOLY[1]

S. Rao Aiyagari
Federal Reserve Bank of Minneapolis

Zvi Eckstein
Tel Aviv University and Boston University

Martin Eichenbaum
Northwestern University and NBER

Abstract

This paper examines the impact of speculative inventories on the rational expectations competitive equilibrium laws of motion of prices, output and sales when firms are confronted with stochastic production and input choices are made prior to the realization of these shocks. The case of a monopolist producer-speculator is also considered. It is shown that a nonnegativity constraint on inventories will, occasionally be binding under both perfect competition and monopoly. Under these circumstances, there exists, for both industry structures, an inventory reservation price which has the property that inventories will be positive when the market price is less than the reservation price. When this is not the case, inventories will be zero. This result is established under general convex production and inventory holding costs.

An example is also provided for calculating the competitive and monopoly equilibria, when corners are taken into account and the shocks to production are independently and identically distributed. The comparative dynamics as well as the impact of industry structure on the means and variances of prices, inventories, production and consumption are analyzed.

1. Introduction

Recently much attention has been focused on the role of inventories in the optimal pricing and output decisions of a monopolist in stochastic environments; see, for example, Blinder (1982), Blinder

---

[1] The authors would like to acknowledge helpful comments by Neil Wallace, Takotoshi Ito, Thomas Sargent, Steven Turnovsky and members of the Labor and Population Workshop at Yale University.

and Fischer (1981), Reagan (1982) and Zabel (1972). Samuelson (1971), Wright and Williams (1982), Kohn (1978), Reagan and Weitzman (1982), and Scheinkman and Schectman (1982), on the other hand, consider the optimal inventory rule in a perfectly competitive market. Kohn considers holding costs of inventories but (like Samuelson) takes production to be exogenous and i.i.d. and notes that it would be interesting to analyze the case in which producers also make non-trivial production choices in light of their beliefs about future prices. Reagan and Weitzman consider the case of constant unit costs of production but no inventory holding costs. This paper examines the impact of speculative inventories on the rational expectations competitive equilibrium laws of motion of prices, output and sales, when firms are confronted with a stochastic production function and input choices are made prior to the realization of these shocks. Both firms and speculators are assumed to maximize the expected present value of their profits. The case of a monopolist producer-speculator is also considered.

It is shown (as, for example, in Kohn, and Reagan and Weitzman) that a nonnegativity constraint on inventories will, occasionally, be binding under both perfect competition and monopoly. Under these circumstances, there exists, for both industry structures, an inventory reservation price which has the property that inventories will be positive when the market price is less than the reservation price. When this is not the case, inventories will be zero. <u>In general, this reservation price is time dependent</u>. However, in the case considered by the existing literature, namely the situation in which the shocks to production are independently and identically distributed, the inventory holding reservation price is time invariant. But, as the paper points out, this will not be the case if the shocks to production are allowed to have a richer stochastic law of motion or if there are costs associated with changing production and/or inventory levels.

The purpose of this paper is to compare and contrast the stochastic behavior of prices, inventories, production and consumption for the competitive monopoly cases as well as to analyze the comparative dynamics of the system. Toward this end, an example is provided for calculating the competitive and monopoly equilibria, when corners are taken into account and the shocks to production are i.i.d.. The comparative dynamics as well as the impact of the industry structure on the means and variances of prices, inventories, output and sales are analyzed. The example illustrates the essentially similar role that inventories play under perfect competition and monopoly. Thus, results obtained by Blinder (1982) and Reagan (1982) for the case of monopoly are shown to carry over to the case of perfect competition. It is also shown that the monopolist exerts a stabilizing influence on the industry. While average prices are higher and average output and sales are lower in the monopoly case, the variances of each of these variables is lower than under perfect competition. This result obtains, at least partly, because of the fact that average inventories are lower but the variance of inventories is higher under monopoly than in a perfectly competitive industry. It is hoped that the method for explicitly computing equilibria when corners are present is of interest, per se, as no such examples are available in the inventory literature. Having such a solution allows for a more detailed characterization of the impact of industry structure on the equilibrium distribution of the endogenous variables considered.

The remainder of this paper is organized as follows. In Section 2 a model is developed of perfectly competitive firms who maximize the present discounted value of an intertemporal profit function subject to output uncertainty. In Section 3, the equilibrium for this industry as well as the monopoly case is analyzed. Section 4 explicitly calculates the equilibrium for a simplified example. Section 5 analyzes the comparative dynamics of the system and contrasts the case of perfect competition with that of a monopoly. Concluding remarks are contained in Section 6.

2. **The Model**

In this section, we characterize the market under consideration by specifying the optimization problems faced by producers and speculators as well as the demand curve for the storable good.[2]

Let there be $N \geq 1$ identical producers. q_t denotes the output of the representative producer at time t. The production function is given by

$$q_t = f(x_{t-1}, \varepsilon_t). \tag{2.1}$$

where x_{t-1} is the quantity of the input X at time t-1 and ε_t is an aggregate shock to time t production. Thus, as in Reagan and Weitzman there is a one-period lag in production. The function $f(x,\varepsilon)$ is assumed to be continuous and for each ε is continuously differentiable, strictly increasing and strictly concave in x. In addition, both $f(x,\varepsilon)$ and $f_1(x,\varepsilon)$ are strictly increasing in ε.

The cost of production at time t is written in terms of the current input decision and is given by the time invariant function $C(x_t)$ which is assumed to be strictly increasing and strictly convex.

Under the above assumptions, the problem of the representative producer, who maximizes the expected value of discounted profits, is[3]

$$\underset{\{x_t\}_{t=0}^{\infty}}{\text{Maximize}} \; E_0 \sum_{t=0}^{\infty} \beta^t \{P_t q_t - C(x_t)\} \tag{2.2}$$

subject to x_{-1} given, (2.1) and the law of motion for the stochastic process $\{P_t\}_{t=0}^{\infty}$. $0 < \beta < 1$ is a discount factor, P_t is the market price of the good at time t. E_t denotes the conditional expectations operator defined on information known at time t. The information set, Ω_t, is assumed to be common to all agents in the model and includes the values of all variables in the model dated t - s, $s \geq 0$.

The first-order conditions for the producer's problem are given by

$$E_t\{\beta P_{t+1} f_1(x_t, \varepsilon_{t+1})\} - C'(x_t) \leq 0 \text{ for all } t = 0, 1, 2, \ldots \tag{2.3}$$

---

[2] The description of the model in this section is heuristic. Rigorous conditions designed to establish existence and uniqueness are postponed to the next section.

[3] For this problem to be well defined, it is sufficient that $\{P_t\}_{t=0}^{\infty}$ be of mean exponential order less than $\beta^{-1/2}$.

We now consider the problem of the representative competitive speculator whose only gains from holding inventories are those that accrue to the speculator by "buying low and selling high". There are M such identical speculators.

Let i_t denote the stock of inventories held by the representative speculator at the end of time t and $d(i_t)$, the cost of holding inventories at time t. We assume that $d(i_t)$ is continuously differentiable, strictly increasing and strictly convex. Notice that by assuming $d'(i_t) \geq 0$ for all $i_t > 0$ we have ruled out a convenience yield argument for holding inventories.[4] The problem of the representative speculator who maximizes the expected discounted value of profits from buying and selling inventories of finished goods is given by

$$\underset{\{i_t\}_{t=0}^{\infty}}{\text{Maximize}} \quad E_0 \sum_{t=0}^{\infty} \beta^t \{P_t(-i_t + i_{t-1}) - d(i_t)\} \qquad (2.4)$$

subject to i_{-1} given, $i_t \geq 0$ for all t and the law of motion for the stochastic process $\{P_t\}_{t=0}^{\infty}$. The first order necessary conditions for this problem are

$$-d'(i_t) + \beta E_t(P_{t+1}) - P_t \leq 0 \text{ for all } t = 0, 1, 2, \ldots \qquad (2.5)$$

We define $X_t = Nx_t$, $I_t = Mi_t$ and $Q_t = Nq_t$ to be the industry-wide quantity of input X used at t, the aggregate stock of inventories in the market at time t and aggregate output at time t. The inverse market demand curve for final consumption of the good is given by

$$P_t = P(D_t) \qquad (2.6)$$

where D_t is the aggregate consumption of the good at time t. To complete the model, we impose the market clearing condition;

$$D_t = Q_t - I_t + I_{t-1}. \qquad (2.7)$$

---

[4] See Brennan (1958) for a discussion of convenience yields.

Before proceeding, we note that if $f(x_{t-1}, \varepsilon_t) \equiv \varepsilon_t$ and ε_t is i.i.d. the present model becomes, in essence, the one studied by Kohn. If $f_1(x_{t-1}, \varepsilon_t) \equiv K$, $c'(x_t) = c$ and $d(I_t) \equiv 0$, where K and c are positive constants, and demand has an additive i.i.d. shock associated with it, we are in the Reagan and Weitzman case. Finally if $f(x_{t-1}, \varepsilon_t) \equiv \varepsilon_t$, $c(x_t) \equiv 0$, $d(i_t) \equiv 0$, ε_t i.i.d. we are in the Samuelson case.

As opposed to the above situation of competitive producers and speculators, the problem of a single monopolist producer/speculator is

$$\text{Maximize}_{\{x_t, i_t\}_{t=0}^{\infty}} \quad E_0 \sum_{t=0}^{\infty} \beta^t \{P(D_t) \cdot D_t - c(x_t) - d(i_t)\} \qquad (2.8)$$

subject to (x_{-1}, i_{-1}) given, and $D_t = f(x_{t-1}, \varepsilon_t) - i_t + i_{t-1}$

The first order necessary conditions for this problem are

$$-d'(i_t) - (MR)_t + \beta E_t (MR)_{t+1} \leq 0 \text{ for all } t = 1, 2, \ldots \qquad (2.9)$$

and

$$\beta E_t \{(MR)_{t+1} f_1(x_t, \varepsilon_{t+1})\} - c'(x_t) \leq 0 \text{ for all } t = 0, 1, 2, \ldots (2.10)$$

where, $(MR)_t = P(D_t) + D_t P'(D_t) =$ marginal revenue at t.

Having considered the problems faced by the individual agents in this economy, corresponding to two industry structures we now turn to the definition of the competitive equilibrium in this market.

<u>Definition</u>

A rational expectations competitive equilibrium is a stochastic process for $\{P_t, D_t, I_t, Q_t, X_t\}_{t=0}^{\infty}$ such that

(i) taking $\{P_t, X_t, D_t, I_t, Q_t\}_{t=0}^{\infty}$ given, $\{i_t, q_t, x_t\}_{t=0}^{\infty}$ solves the problems of the representative producer and speculator,

(ii) the sequence $\{P_t, D_t, I_t, X_t, Q_t\}_{t=0}^{\infty}$ which agents take as given when solving their optimization problem is implied by the solution to those problems so that

$i_t = I_t/M$, $q_t = Q_t/N$, $x_t = X_t/N$, for all t,
(iii) the goods market clears, so that

$$P_t = P(D_t) = P(Q_t - I_t + I_{t-1}), \text{ for all } t \geq 0.$$

We now turn to the properties of this equilibrium. The equilibrium for the monopoly case will be derived as a special case of the competitive equilibrium.

3. Existence and Uniqueness Properties of the Rational Expectations Equilibrium

In this section, we establish the existence of a unique competitive equilibrium and characterize some of its properties. In addition, we analyze the case of a single monopolistic producer/speculator. We first analyze the case of $\{\varepsilon_t\}_{t=0}^{\infty}$ being i.i.d. We then allow $\{\varepsilon_t\}_{t=0}^{\infty}$ to follow a general finite order autoregressive scheme.

To begin with, we assume that ε_t is i.i.d. and bounded so that

$$-\infty < \varepsilon_t \leq \varepsilon_t \leq \varepsilon_b < \infty \text{ for all t.} \qquad (3.1)$$

In addition, we assume the following regularity conditions.

$$P(0) < \infty \qquad (3.2)$$

There exists an X* such that

$$C'(X^*/N) > \beta P(0) f_1(X^*/N, \varepsilon_b) \qquad (3.3)$$

$$d'(0) = 0 \qquad (3.4a)$$

and there exists an I* such that

$$d'(I^*/M) > \beta P(0) \qquad (3.4b)$$

Let $y^* = I^* + Nf(X^*/N, \varepsilon_b)$ and $P_m = P(y^*)$.

$$0 < P_m < \beta P[Nf(X^*/N, \varepsilon_b)] \tag{3.5}$$

$$C'(0) < \beta P_m f_1(0, \varepsilon_a) \tag{3.6}$$

Because the inverse demand function has a negative slope, (3.2) guarantees that the equilibrium price distribution is bounded from above. (3.3) asserts that there exists a level of input, X^*/N, such that the marginal cost of the input at that level exceeds the discounted value of the marginal product of the input evaluated at the maximum possible price and the best possible production shock. Thus, (3.3) guarantees that, the input into the production process and, therefore, output is bounded from above. The first part of (3.4) assumes that the marginal inventory holding costs associated with zero inventories is zero. This implies that if inventories are zero and prices are expected to rise, then positive inventories will be held. The second part of (3.4) assumes that there exists a level of inventories, I^*/M, such that the marginal inventory holding costs exceeds the discounted value of the maximum price. This is sufficient to guarantee that inventories are bounded from above. To interpret (3.5) and (3.6) we note that y^* denotes the maximum possible consumption of the good. (3.5) then says that the price corresponding to this level as given by the demand curve is positive but less than the discounted value of the price corresponding to sales equal to the maximum output level. Since the former price is a lower bound on the price distribution, (3.5) implies that if the lower bound on prices is attained, positive inventories will be held. If (3.5) was not true, it is quite possible that, despite the storability of the good, inventories would never be held. Finally (3.6) says that the marginal cost of the input associated with a zero input level is less than the discounted lowest possible value of the marginal product of the input. Given this condition input use and hence output will always be positive.

Given the regularity conditions, (3.2) - (3.6), consider the following problem

Maximize

$$E_0 \sum_{t=0}^{\infty} \beta^t [\int_0^{D_t} P(Z)dZ - NC(X_t/N) - Md(I_t/M)] \qquad (3.7)$$

subject to

$$D_t + I_t \leq I_{t-1} + Q_t$$

$$Q_t = Nf(X_{t-1}/N, \varepsilon_t)$$

$$D_t \geq 0, \quad 0 \leq X_t \leq X^*, \quad 0 \leq I_t \leq I^*$$

X_{-1}, I_{-1}, ε_0 given.

Our objective is to show that problem (3.7) has a unique solution which is also the solution to the rational expectations competitive equilibrium defined in Section 2. Notice that

$$U(D_t, X_t, I_t) = \int_0^{D_t} P(Z)dZ - NC(X_t/N) - Md(I_t/M) \qquad (3.8)$$

is simply the area under the market demand curve for final consumption of the good minus the costs of production and holding inventories. Hence, problem (3.7) may be interpreted as the social planning problem of maximizing the expected discounted value of producer plus consumer surplus. Thus, to establish that the competitive equilibrium solves problem (3.7) is to show that it is optimal with respect to this welfare criterion. This is simply the well-known result regarding the optimality of competitive equilibria (see for example, Samuelson, Kohn, as well as Lucas and Prescott (1971) in a different context).

We proceed with the analysis by noting that the function U defined by (3.8) is continuously differentiable and strictly concave. As a consequence, problem (3.7) is a standard concave discounted dynamic programming problem which may be solved as follows:

Let y_t denote the maximum possible consumption of the good at time t,

$$y_t = I_{t-1} + Nf(X_{t-1}/N, \varepsilon_t) \qquad (3.9)$$

The set S is defined as

$$S = \{ V \mid V: [0, y^*] \to R: \} \quad (3.10)$$

where V is continuous, non-decreasing and concave. The mapping T: S → S is defined by

$$TV(y_t) = \max_{D_t, X_t, I_t} \left\{ \begin{array}{l} U(D_t, X_t, I_t) + \beta E_t V(I_t + Nf(X_t/N, \varepsilon_{t+1})) \\ \text{subject to } D_t + I_t \leq y_t \\ 0 \leq D_t, \; 0 \leq X_t \leq X^*, \; 0 \leq I_t \leq I^* \end{array} \right\} \quad (3.11)$$

It is straightforward to establish that S is complete in the supremum metric and that T is a contraction. Hence, there is a unique $V \in S$ satisfying

$$V(y_t) \equiv TV(y_t) \quad (3.12)$$

as well as unique and continuous optimal policy functions

$$D_t = D(y_t) \quad (3.13a)$$

$$X_t = X(y_t) \quad (3.13b)$$

$$I_t = I(y_t) \quad (3.13c)$$

$$y_t = I_{t-1} + Nf(X_{t-1}/N, \varepsilon_t) \quad (3.13d)$$

which attain the maximum on the right hand side of (3.11) for each y_t. It is also seen that $V(\cdot)$ is strictly increasing and strictly concave because U is strictly increasing in D_t and is strictly concave.

Moreover, proceeding as in Sargent (1979), one may show that V is, in fact, continuously differentiable.[5]

Using the differentiability of $V(\cdot)$, one may rewrite (3.11) and (3.12) by forming the Lagrangian:

$$V(y_t) = \max_{D_t,X_t,I_t,\lambda_t} \left\{ U(D_t,X_t,I_t) + \beta E_t V(I_t + Nf(X_t/N,\varepsilon_{t+1}) + \lambda_t(y_t - D_t - I_t) \right\} \quad (3.14)$$

The necessary (and sufficient) conditions for a maximum are:

$$P(D_t) - \lambda_t = 0 \quad (3.15)$$

$$-C'(X_t/N) + \beta E_t[V'(I_t + Nf(X_t/N,\varepsilon_{t+1}))f_1(X_t/N,\varepsilon_{t+1})] = 0 \quad (3.16)$$

$$-d'(I_t/M) + \beta E_t[V'(I_t + Nf(X_t/N,\varepsilon_{t+1}))] - \lambda_t \leq 0, = 0 \text{ if } I_t > 0 \quad (3.17)$$

$$y_t = D_t + I_t \quad (3.18)$$

Further, from (3.14) we have that

$$\lambda_t = V'(y_t) \quad (3.19)$$

The optimal policy functions (3.13) are the solutions of (3.15) through (3.19).

Notice that since $\lambda_{t+1} = V'(y_{t+1}) = V'(I_t + Nf(X_t/N, \varepsilon_{t+1}))$, (3.15) through (3.18) can be written as

$$\lambda_t = P(D_t) \quad (3.20a)$$

---

[5]The essential steps in proving this are:
(i) show that if $w(\cdot)$ is continuously differentiable then $Tw(\cdot)$ is continuously differentiable.
(ii) consider the sequence of functions in S: $V_0 = 0$, $V_n = TV_{n-1}$, n = 1,2,3,... One demonstrates that the sequence of derivative functions $\{V'_n\}_{n=0}^{\infty}$ converges uniformly. Since $\{V_n\} \to V$, it is then straightforward to show that V is continuously differentiable and $\{V'_n\} \to V'$.

$$C'(X_t/N) = \beta E_t[\lambda_{t+1} f_1(X_t/N, \varepsilon_{t+1})] \tag{3.20b}$$

$$d'(I_t/M) \geq E_t[\beta\lambda_{t+1} - \lambda_t] \text{ (with equality if } I_t > 0) \tag{3.20c}$$

$$D_t + I_t = y_t = I_{t-1} + Nf(X_{t-1}/N, \varepsilon_t) \tag{3.20d}$$

By using the fact that $X_t = Nx_t$, $I_t = Mi_t$, $Q_t = Nq_t = Nf(x_{t-1}, \varepsilon_t)$, system (3.20) is seen to be identical with the rational expectations competitive equilibrium defined in Section 2. Hence, the existence of a unique equilibrium has been established.

We now establish the properties of the optimal policy functions (3.13). In appendix A it is shown that the constraints $0 \leq X_t \leq X^*$ and $I_t \leq I^*$ are nonbinding.
(i) From (3.13) and (3.15), since V is strictly concave and $P(\cdot)$ is strictly decreasing, we have that $D(y_t)$ is strictly increasing.
(ii) $X(y_t)$ is non-increasing and $I(y_t)$ is non-decreasing. For, suppose to the contrary that $I(y_t)$ falls as y_t rises in some range. Then (3.17) holds with equality over this range so that $X(y_t)$ must also fall. But then (3.16) cannot hold. Hence $I(y_t)$ must be nondecreasing which implies from (3.16) that $X(y_t)$ must be non-increasing.

We now state and prove the following proposition which establishes, for the model under consideration, that industry-wide inventories exhibit a time-invariant reservation price property.

Proposition 2

There exists a time invariant price, $\hat{P}$, such that

$$P_t \geq \hat{P} \Rightarrow I_t = 0 \tag{3.21a}$$

$$P_t < \hat{P} \Rightarrow I_t > 0 \tag{3.21b}$$

Proof

We first show that there exist $\hat{X}$ and $\hat{y}$ which satisfy:

$$C'(\frac{\hat{X}}{N}) = \beta E_t[V'(Nf(\frac{\hat{X}}{N}, \varepsilon_{t+1}))f_1(\hat{X}_t/N, \varepsilon_{t+1})] \tag{3.22}$$

$$V'(\hat{y}) = \beta E_t[V'(Nf(\hat{X}/N, \varepsilon_{t+1}))] \tag{3.23}$$

From (3.6) we have:

$$C'(0) < \beta P_m f_1(0, \varepsilon_a) \stackrel{<}{=} \beta P_m E_t f_1(0, \varepsilon_{t+1})$$
$$\leq \beta E_t[V'(Nf(0, \varepsilon_{t+1})) f_1(0, \varepsilon_{t+1})] \tag{3.24}$$

From (3.3) we get:

$$C'(\frac{X^*}{N}) > \beta P(0) f_1(X^*/N, \varepsilon_b) \geq \beta P(0) E_t f_1(X^*/N, \varepsilon_{t+1})$$
$$\geq \beta E_t[V'(Nf(X^*/N, \varepsilon_{t+1})) f_1(X^*/N, \varepsilon_{t+1})] \tag{3.25}$$

From (3.24) and (3.25) it is immediate that there is a unique $\hat{X}$, $0 < \tilde{X} < X^*$ solving (3.22).

Next we show that if $y_t = y^*$ then $I(y^*) > 0$. Suppose to the contrary that $I(y^*) = 0$. Then $D(y^*) = y^*$ and from (3.15) and (3.19)

$$V'(y^*) = P(y^*) = P_m.$$

From (3.16) and (3.22) we have $X(y^*) = \hat{X}$ and from (3.17):

$$\beta E_t V'[Nf(\hat{X}/N, \varepsilon_{t+1})] \leq \lambda_t = V'(y^*) = P_m$$

Further, $V'(y_t) = P(D_t) \geq P(y_t)$ since $D_t \leq y_t$. Hence:

$$\beta P(Nf(X^*/N, \varepsilon_b)) \leq \beta E_t[P(Nf(\hat{X}/N, \varepsilon_{t+1}))]$$
$$\leq \beta E_t[V'(Nf(\hat{X}/N, \varepsilon_{t+1}))] \leq P_m$$

which contradicts assumption (3.5). Hence $I(y^*) > 0$. Further, $I(y_t)$ cannot be strictly positive for all $y_t > 0$ because then (3.17) holds at equality for all $y_t > 0$ and hence, by continuity, at $y_t = 0$. This implies $I(0) = 0$ and from (3.16), $X(0) = \hat{X}$. (3.17) becomes:

$$V'(0) = \beta E_t V'(Nf(\hat{X}/N, \varepsilon_{t+1})) \leq \beta P(0)$$

which is a contradiction since $V'(0) = P(0)$.

It now follows that:

$$\hat{y} = \sup\{y \mid I(y) = 0\}$$

satisfies $0 < \hat{y} < y^*$. This $\hat{y}$ also satisfies (3.23) and is unique. By defining $\hat{D} = D(\hat{y})$ and $\hat{P} = P(\hat{D})$ it is immediately evident that

$$P_t \geq \hat{P} \Rightarrow I_t = 0$$

and

$$P_t < \hat{P} \Rightarrow I > 0.$$

The intuition behind this result is quite straightforward. Given the i.i.d. specifications of the ε_t process, if the current market price is high (i.e., ε_t and hence current production is low), then the (discounted) expected price next period will almost surely be lower and inventories will not be held. On the other hand, if the current market price is "low", then the (discounted) expected price next period will almost surely be higher and it pays to hold some inventories. Viewed in this way, the time invariant nature of the reservation price is seen to depend crucially upon the assumption that ε_t is i.i.d. However, the above analysis, per se, can easily be extended to allow for serial correlation in ε_t.[6]

To illustrate, suppose that ε_t follows a z'th order autoregressive progress, for some finite z, i.e.,

$$\text{Prob}[\varepsilon_t \leq \varepsilon \mid \varepsilon_{t-j}, j \geq 1] = \text{Prob}[\varepsilon_t \leq \varepsilon \mid \varepsilon_{t-1}, \varepsilon_{t-2}, \ldots, \varepsilon_{t-z}]$$

---

[6] Samuelson (1971) considers a similar model of speculative behavior where inventories are subject to "shrinkage" but there are no costs to holding inventories. In addition, production is given as an exogenous stochastic process which is independent over time. His result regarding the existence of a reservation price is similar to ours.

Let $\alpha_t = (\varepsilon_t, \varepsilon_{t-1}, \ldots, \varepsilon_{t-z+1})$ and rewrite (3.12) as

$$\tilde{V}(y_t, \alpha_t) \equiv \max_{D_t, X_t, I_t} \left\{ \begin{array}{l} U(D_t, X_t, I_t) + \beta E_t[V(I_t + NF(X_t/N, \varepsilon_{t+1}), \alpha_{t+1})] \\ \text{subject to:} \\ 0 \leq D_t, 0 \leq X_t \leq X^*, 0 \leq I_t \leq I^*, D_t + I_t \leq y_t \end{array} \right\}$$

where the expectation $E_t(\cdot)$ on the r.h.s. is taken conditional on α_t. All of the steps in the previous analysis can now be carried out in exactly the same way with the replacement of $V'(\cdot)$ by $V_1(\cdot,\cdot)$. The arguments used before would be interpreted as corresponding to a fixed value of α_t. Obviously, the reservation price characteristic of inventory holdings would be modified. In the i.i.d. case the reservation price was a constant over time whereas in the present case, for each t the reservation price depends on α_t, i.e., on current and past realizations of ε_t.

In concluding this section, we briefly discuss the case of a single monopolist producer/speculator. This is done to indicate that the above analysis can accommodate this industry structure and that the reservation price property is not peculiar to the case of perfect competition. To do this we set M = N = 1 and define the revenue function $R(D_t) = D_t P(D_t)$ which replaces the area under the demand curve $\int_0^{D_t} P(Z) \, dZ$ used previously. Having done this, (3.7) becomes identical to (2.8) which defined the objective function of the monopolist. We need to assume that the marginal revenue function is well-defined and satisfies assumptions similar to those imposed on the inverse demand curve for the competitive case. Then the existence of a unique monopoly equilibrium and of the reservation price property of inventory holdings are guaranteed.

4. The Equilibrium for a Linear Model

In Sections 2 and 3 we saw that speculative inventories cannot always be nonnegative if they are unconstrained. Moreover, the imposition of a nonnegativity constraint implies that the calculation of an equilibrium leads to a nontrivial problem with corners, in that the constraint will occasionally be binding. In this section, we exhibit an explicit solution for the rational expectations competitive equilibrium as well as the monopoly problem. The reason for pursuing an explicit solution is to analyze the properties of equilibrium inventories, as well as to contrast the behavior of the endogenous variables under the two industry structures considered. In addition, we develop the "comparative dynamics" of the system.

The example considered reproduces the models considered by Muth (1961) and Turnovsky (1978) by specifying explicit optimization problems for producers and speculators.

The production function is specified as

$$q_t = f(x_{t-1}, \varepsilon_t) = \alpha_1 x_{t-1} + \frac{\varepsilon_t}{N}, \qquad \alpha_1 > 0 \tag{4.1}$$

The costs of production are given by

$$C(x_t) = \frac{\delta}{2} x_t^2 + R x_t \ ; \quad \delta, R > 0 \tag{4.2}$$

The costs of holding inventories are

$$d(I_t) = \frac{1}{2d} I_t^2 \ ; \qquad d > 0 \tag{4.3}$$

The demand function is given by

$$D_t = A - a P_t, \qquad A, a > 0 \tag{4.4}$$

Given this specification, the aggregate supply function is given by

$$Q_t = b E_{t-1} P_t - C + \varepsilon_t \tag{4.5}$$

where $b = N\beta\alpha_1^2/\delta$ and $C = N\alpha_1 R/\delta$. In what follows, we assume that M=N=1. Obviously, the notion of a competitive equilibrium with a single producer-speculator is very strained. However, this is assumed here chiefly to facilitate comparisons of the competitive and monopoly equilibria. The first order conditions for the problems faced by speculators become

$$I_t = \begin{cases} d[\beta E_t P_{t+1} - P_t] & \text{if } \beta E_t P_{t+1} \geq P_t \\ 0 & \text{otherwise} \end{cases} \quad (4.6)$$

In addition, we have the market clearing condition

$$D_t + I_t = Q_t + I_{t-1} \quad (4.7)$$

In order to obtain an explicit solution for the model, we assume that the shock to production, ε_t, has the following i.i.d. probability distribution:

$$\varepsilon_t = \begin{cases} \varepsilon_1 \text{ with probability } \pi_1 \text{ for all } t \\ \varepsilon_2 \text{ with probability } \pi_2 \text{ for all } t \end{cases} \quad (4.8)$$

where $\pi_1 + \pi_2 = 1$, $\varepsilon_1 > \varepsilon_2$, and $\pi_1 \varepsilon_1 + \pi_2 \varepsilon_2 = 0$. It is clear that, in the absence of inventories, the sequence of equilibrium prices would be i.i.d. However, with inventories this will not be the case. This may be contrasted with Reagan and Weitzman who obtain an i.i.d. equilibrium price distribution because of constant unit costs of production. Finally, before proceeding, we note that three cases of the above specification, $(\alpha_1 = 0, d = \infty)$, $(\alpha_1 = 0, d < \infty)$ and $(\delta = 0, d = \infty)$, correspond to special cases of the models considered by Samuelson, Kohn, and Reagan and Weitzman, respectively.

From Proposition 2 we know that there exists a time invariant inventory holding reservation price, P^*, such that $I_t > 0$ if $P_t < P^*$ and $I_t = 0$ if $P_t \geq P^*$. As such, we begin with the following guess regarding the equilibrium price process.

Notice that in the above figure, the functions ℓ^1 and ℓ^2 are parameterized by the realization of the ε_t process and have a kink at P^*. This leads us to propose the following solution:

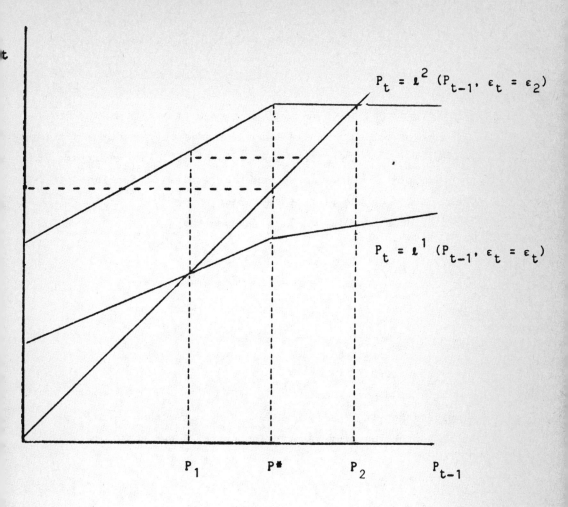

Figure 1

$$P_t = \lambda_1(\varepsilon_t)P_{t-1} + \mu_1(\varepsilon_t), \quad P_1 \leq P_{t-1} \leq P^* \tag{4.9}$$

$$P_t = \lambda_2(\varepsilon_t)P_{t-1} + \mu_2(\varepsilon_t), \quad P^* \leq P_{t-1} \leq P_2. \tag{4.10}$$

Figure 1 implies that the equilibrium price will remain in the interval $[P_1, P_2]$ with probability one. Hence, we need only consider the range $[P_1, P_2]$ for P_{t-1}. It is clear that we are implicitly searching for a stationary solution. However, given the results of Section 3 this is not a restriction on the proposed solution.

As is evident from Figure 1, we assume that

$$\lambda_1(\varepsilon_2)P_1 + \mu_1(\varepsilon_1) > P^* \tag{4.11}$$

which implies that, if $P_1 \leq P_{t-1} \leq P^*$ and $\varepsilon_t = \varepsilon_2$, then $P^* \leq P_t \leq P_2$. and

$$\lambda_2(\varepsilon_1)P_2 + \mu_2(\varepsilon_1) < P^* \tag{4.12}$$

which implies that if $P^* \leq P_{t-1} \leq P_2$ and $\varepsilon_t = \varepsilon_1$, then $P_1 \leq P_t \leq P^*$.[7]

The inventory holding reservation price P^* has the property that

$$P_1 \leq P_t \leq P^* \Rightarrow I_t \geq 0 \tag{4.13}$$

$$P^* \leq P_t \leq P_2 \Rightarrow I_t = 0.$$

We are able to compute the equilibrium under the above assumptions. As some of the details are tedious, the procedures and computations are given in Appendix B. In addition to obtaining expressions for $\{\lambda_1(\varepsilon_t), \lambda_2(\varepsilon_t), \mu_1(\varepsilon_t), \mu_2(\varepsilon_t), \varepsilon_t = \varepsilon_1, \varepsilon_2\}$, it is shown that the reservation holding price for inventories is given by

$$P^* = \frac{\beta \bar{\mu}_1}{1 - \beta \bar{\lambda}_1} \tag{4.14}$$

---

[7] Equations (4.11) and (4.12) impose restrictions on the basic parameters of the model.

where P* satisfies (4.13) and the "-" symbol stands for the expected value so that, for example $\bar{\lambda}_1 = \pi_1 \lambda_1(\varepsilon_1) + \pi_2 \lambda_1(\varepsilon_2)$.

Distribution of Inventories

Because P_t is serially correlated we expect I_t to be serially correlated. This turns out to be the case and the behavior of inventories can be represented as in Figure 2.

Let $d[\beta\bar{\mu}_1 - P_1(1 - \beta\bar{\lambda}_1)] = I^*$ and $\lambda_1(\varepsilon_1) = \alpha$. Then inventories will take on the values $\{0, (1 - \alpha)I^*, (1 - \alpha^2)I^*..., (1 - \alpha^N)I^*, ...\}$ with the corresponding probabilities $\{\pi_2, \pi_2\pi_1, \pi_2\pi_1^2, ..., \pi_2\pi_1^N, ...\}$. It is easy to show that the unconditional expected value of inventories is equal to

$$E[I_t] = \frac{\pi_1(1 - \alpha)I^*}{1 - \pi_1\alpha}.$$

Furthermore, the probability of observing zero as opposed to positive inventories is π_2, the probability of a negative shock to output. Hence, if π_2 is small, inventories will be positive most of the time.

An interesting feature of the solution, as depicted in Figure 2, is that starting from zero inventories, a succession of "good" shocks ($\varepsilon_t = \varepsilon_1$) results in a build up of inventoreis (at a decreasing rate) and then a complete decumulation of stocks when a "bad" shock is realized.

It can also be shown that a larger value of d, i.e., a lower cost of holding inventories, results in a higher expected value of inventories.

By using (4.14) we may write the optimal inventory decision rule as

$$I_t = \begin{cases} d(1 - \beta\bar{\lambda}_1)[P^* - P_t] & \text{for } P_t \leq P^* \\ 0 & \text{for } P_t > P^* \end{cases} \qquad (4.15)$$

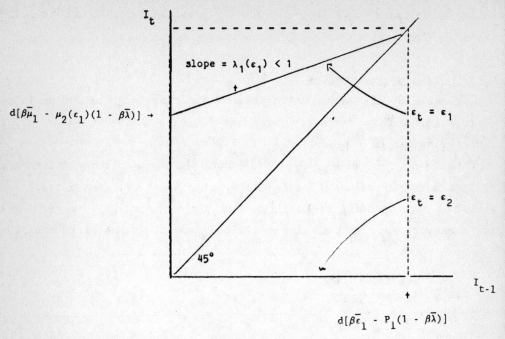

Figure 2

(4.15) emphasizes the role of the reservation price P* and demonstrates that the level of inventories, when positive, is strictly proportional to the difference between P* and P_t.

5. Comparative Dynamics, Comparison of Monopoly and Competition

An important advantage of our analytical solution is that one can, in principle, analyze the effects of changes in the underlying parameters of the model on the first and second moments of the distributions of prices, inventories, consumption and production. However, even for the simple example considered here, these calculations are difficult due to the presence of corners. As such, we simulated the model by generating a sequence of three hundred random numbers for the shocks to production, ε_t. We then calculated the first and second moments of the variables for various values of the underlying parameters.

Using the initial numerical example, we analyze the effects of changes in the underlying parameters. In addition, the competitive equilibrium is then compared to the monopoly solution.

Let $\beta = .9$, $d = 10.0$, $b = 5.0$, $A = 12.0$, $a = 5.0$, $C = 2.0$,
$\varepsilon_1 = 1.0$, $\pi_1 = .8$, $\varepsilon_2 = -4.0$, $\pi_2 = .2$.

For the example, in Section 4 we obtain the following equilibrium sequences

$$P_t = \begin{cases} .351\ P_{t-1} + .784 & \text{for } P_1 \leq P_{t-1} \leq P^*,\ \varepsilon_t = \varepsilon_1 \\ .778\ P_{t-1} + 1.182 & \text{for } P_1 \leq P_{t-1} \leq P^*,\ \varepsilon_t = \varepsilon_2 \end{cases} \quad (5.1a)$$

$$P_t = \begin{cases} 1.234 & \text{for } P^* \leq P_{t-1} \leq P_2,\ \varepsilon_t = \varepsilon_1 \\ 1.177 & \text{for } P^* \leq P_{t-1} \leq P_2,\ \varepsilon_t = \varepsilon_2 \end{cases} \quad (5.1b)$$

$$I_t = \begin{cases} -6.071\ P_t + 7.773 & P_t \leq 1.28 = P^* \\ 0 & P_t \leq 1.28 = P^* \end{cases} \quad (5.1c)$$

$$Q_t = 5.0\ E_{t-1}(P_t) - 2.0 + \varepsilon_t. \quad (5.1d)$$

Using the above as a benchmark, we calculated the means and variances of prices, inventories, consumption and production for a sequence of values of a, b, d and β. We changed the value of each of the parameters, holding the other constant at their benchmark levles.

The results of these experiments are reported in Table 1. Recall that: an increase in <u>a</u> implies a decrease in the slope of the demand, an increase in <u>b</u> implies a decrease in the slope of the supply curve, an increase in <u>d</u> is equivalent to a decrease in the cost of holding inventories, and an increase in β is equivalent to a decrease in the interest rate.

Table 1: Comparative Dynamics

an increase in	Price		Inventories		Consumption		Production	
	mean	variance	mean	variance	mean	variance	mean	variance
a	(-)	(-)	(-)	(-)	(-)	(+)	(-)	(-)
b	(-)	(-)	(+)	(+)	(+)	(-)	(+)	(+)
d	(-)	(-)	(+)	(+)	(+)	(-)	(+)	(+)
β	(+)	(-)	(+)	(+)	(-)	(-)	(-)	(-)

It is interesting to note that if one interprets the holding costs of inventories as being inversely related to the durability of the good, then Table 1 has the implication that the variance of production is positivily related to the durability of the good. This is, of course, consistent with the fact, documented by Hodrick and Prescott (1978) and discussed by Lucas (1977) among others, that the production of producer and consumer duarbles has a larger variance that does the production of non-durables.

Observe that the variances of inventories and consumption always move in opposite directions. This clearly reflects the fundamental role of inventories in this model -- the stabilization of consumption. For example, when the cost of holding inventories (1/d) goes down, the means of production, consumption and inventories while prices fall on average. However, while the variance of inventories and production increase, the variances of consumption and prices fall. Similarly, an increase in the discount rate stabilizes consumption and production (while decreasing their means), but increases the variance of inventories.

Finally, the solution to the model when the nonnegativity constraint not imposed is:[8]

$$P_t = .850 + .392 P_{t-1} - .087 \varepsilon_t$$

$$I_t = -6.466 P_t + 7.652$$

$$D_t = 12.0 - 5.0 P_t$$

$$Q_t = 2.25 + 1.963 P_{t-1} + \varepsilon_t.$$

The variance of prices for the above two models are as follows:
The model of this paper
$$\sigma_p^2 = 0.096$$
The model where inventories can be negative
$$\sigma_p^2 = 0.036$$
Finally, if no inventories are allowed, the variance of prices is given by
$$\sigma_p^2 = 0.16$$

It should not come as a surprise that the variance of prices is largest when inventories are outlawed, (0.16), followed by the case in which inventories are allowed but are constrained to be non-negative (0.096) and that the variance of prices is smallest when the nonnegativity constraint on inventories is not imposed (0.036). The effect of inventories is clearly to reduce price variance. This result is very much in the spirit of Reagan (1982) who shows that when firms are stocked out, ($I_t = 0$), shocks to demand have more of an impact on prices than when $I_t > 0$.

Finally, we note that the solution for inventories in the constrained case is quite close to the unconstrained case. However the

---

[8] In some papers, this is justified by interpreting negative inventories as backlogs.

equilibrium price processes are very different. This results in inventories having a negative mean for the latter model, while in the former model, inventories are, by construction, never negative. These overall differences seem to offer support to recent empirical work by Reagan and Sheehan (1982) who point out the marked differences in the statistical properties of (U.S.) manufacturing inventories of finished goods and backlogs of unfilled orders. Their arguments suggest that the latter should not be treated as negative quantities of the former variable. The results in this section suggest that pursuing their strategy may also lead to important different theoretical implications for the analysis of inventories and backlogs.

We consider the monopolist producer/speculator case. As we noted in Section 3, this case is quite similar to that of perfect competition. In fact, it is straightforward to demonstrate that, given a linear demand function, the monopoly solution for output and inventories may be obtained by replacing the parameters (A,a) by (A/2, a/2) in the social planning problem whose solution yields the competitive equilibrium. With this specification P_t is simply interpreted as the marginal revenue accruing to the monopolist at time t, (MR_t) so that there is now a time invariant reservation marginal revenue, MR*, such that

$$MR_t < MR^* \Rightarrow I_t > 0, \text{ and}$$

$$MR_t \geq MR^* \Rightarrow I_t = 0.$$

Using the same parameter specification as in the competitive case, one obtains the following solutions

$$MR_t = \begin{cases} .358 MR_{t-1} + .485 & \text{for } MR_1 \leq MR_{t-1} \leq MR^*, \ \varepsilon_t = \varepsilon_1 \\ 1.130 MR_{t-1} + 1.240 & \text{for } MR_1 \leq MR_{t-1} \leq MR^*, \ \varepsilon_t = \varepsilon_2 \end{cases}$$

$$MR_t = \begin{cases} .864 & \text{for } MR^* \leq MR_{t-1} \leq MR_2, \ \varepsilon_t = \varepsilon_1 \\ 2.44 & \text{for } MR^* \leq MR_{t-1} \leq MR_2, \ \varepsilon_t = \varepsilon_2 \end{cases}$$

$$I_t = \begin{cases} 5.387 MR_{t-1} + 5.72, & MR_t \leq 1.06 = MR* \\ 0, & MR_t \geq 1.06 = MR* \end{cases}$$

We then calculated the mean and variance of market prices, inventories, production and consumption for the competitive and monopoly cases. The results, which are presented in Table 2, are based upon a random sequence (as before $\pi_1 = .8$ and $\pi_2 = .2$) of three hundred value of ε_t.

Table 2: Comparison between competition and monopoly

	Price		Inventories		Consumption		Production	
	mean	variance	mean	variance	mean	variance	mean	variance
Competition	1.4	.16	3.5	.24	4.9	3.83	4.9	5.03
Monopoly	1.7	.04	1.19	.50	3.3	2.19	3.3	4.4

As one might expect, the mean of prices is higher, while average production and sales are lower under monopoly than under perfect competition. Moreover, on average, the monopolist holds <u>less</u> inventories. But because the variance of inventories is <u>higher</u>, the monopolist reduces the variance of prices compared to the case of perfect competition. At the same time, the variances of consumption and production are also lower under monopoly. Thus the monopolist stabilizes all the variables in the model, with the exception of inventories.

At the same time, the above example, along with the fact that the equilibrium obtaining under the monopolist is suboptimal (see Section 3), are supportive of the general claim that stability of prices or consumption per se, is not necessarily a desirable one from a welfare point of view.

6. Concluding Remarks

A model is developed of an industry which produces and sells a storable good. The rational expectations competitive equilibrium is obtained by solving a well-defined social planning problem. The monopoly case is also considered. Both industry structures exhibit an optimal inventory holding reservation price. The properties of this price are tied to the properties of shocks to output. A specialized example is then considered which exhibits the stabilizing role of the monopolist on prices, output, and sales. This may be explained, at least in part, by the fact that the monopolist holds less inventories on average but varies them more in order to accommodate shocks to production.

Appendix A

Here we show that the constraints $0 \leq X_t \leq X^*$ and $I_t \leq I_t^*$ will be non-binding. We have that

$$D_t + I_t = y_t = I_{t-1} + Nf(\frac{X_{t-1}}{N}, \varepsilon_t) \leq y^*$$

Hence, $D_t \leq y^*$. Then, (3.5), (3.15) and (3.19) imply that

$$P_m \leq V'(y_t) \leq P(0) \tag{A.1}$$

Hence from (3.3) and (3.16) we have that

$$\beta P_m E_t [f_1(\frac{X_t}{N}, \varepsilon_{t+1})] \leq C'(\frac{X_t}{N}) \leq \beta P(0) E_t[f_1(\frac{X_t}{N}, \varepsilon_{t+1})]$$

and by using (3.1), we obtain that

$$\beta P_m f_1(\frac{X_t}{N}, \varepsilon_a) \leq C'(\frac{X_t}{N}) \leq \beta P(0) f_1(\frac{X_t}{N}, \varepsilon_b)$$

Using (3.3) and (3.6) we conclude that

$$0 < X_t < X^*$$

Secondly, from (3.17), if $I_t > 0$ we have

$$d'(I_t/M) = \beta E_t[V'(I_t + Nf(\frac{X_t}{N}, \varepsilon_{t+1}))] - \lambda_t \leq \beta P(0)$$

Using (3.4b) we conclude that

$$I_t < I^*.$$

Finally, the constraint $y_t \geq D_t + I_t$ is easily seen to be binding by virtue of (3.5).

Appendix B

In this appendix, we discuss the computation of the equilibrium discussed in Section 4.

Given that specification there exist four possible cases for the market clearing condition (4.7) corresponding to the two ranges for P_{t-1} and the two realizations for ε_t.

Case 1

$P_1 \leq P_{t-1} \leq P^*$, $\varepsilon_t = \varepsilon_1$ which implies $P_1 \leq P_t \leq P^*$ and $I_{t-1} \geq 0$ and $I_t \geq 0$.

By taking conditional expectations on the relevant price processes and substituting into the supply and inventory demand functions, (4.5) and (4.6) respectively, market clearing requires that

$$A - aP_t + d[\beta(\bar{\lambda}_1 P_t + \bar{\mu}_1) - P_t] = b[\bar{\lambda}_1 P_{t-1} + \bar{\mu}_1] - C + \varepsilon_1$$

$$+ d[\beta(\bar{\lambda}_1 P_{t-1} + \bar{\mu}_1) - P_{t-1}] \qquad (B.1)$$

where the "-" symbol stands for the expected value. Substituting (4.10) with $\varepsilon_t = \varepsilon_1$ into the above expression and equating the coefficients of P_{t-1}, as well as the constant terms, we obtain,

$$\lambda_1(\varepsilon_1)[a + d(1 - \beta\bar{\lambda})] = d(1 - \beta\bar{\lambda}_1) - b\bar{\lambda}_1 \qquad (B.2)$$

$$\mu_1(\varepsilon_1)[a + d(1 - \beta\bar{\lambda}_1)] = A + C - \varepsilon_1 - b\bar{\mu}_1 \qquad (B.3)$$

Case 2

$P_1 \leq P_{t-1} \leq P^*$, $\varepsilon_t = \varepsilon_2$

which implies that $P^* \leq P_t \leq P_2$, and $I_{t-1} \geq 0$ and $I_t = 0$.

Proceeding as above, we obtain,

$$a\lambda_1(\varepsilon_2) = d(1 - \beta\bar{\lambda}) - b\bar{\lambda} \qquad (B.4)$$

$$a\mu_1(\varepsilon_2) = A + C - \varepsilon_2 - (b + \beta d)\bar{\mu}_1 \qquad (B.5)$$

Case 3

$$P^* \leq P_{t-1} \leq P_2, \quad \varepsilon_t = \varepsilon_1$$

which implies that $P_1 \leq P_t \leq P^*$ and $I_{t-1} = 0$ and $I_t \geq 0$.

Again, proceeding as above, we obtain,

$$\lambda_2(\varepsilon_1)[a + d(1 - \beta\bar{\lambda})] = -b\bar{\lambda}_2 \tag{B.6}$$

$$\mu_2(\varepsilon_1)[a + d(1 - \beta\bar{\lambda}_1)] = A + C - \varepsilon_1 - b\bar{\mu}_2 + \beta d\bar{\mu}_1 \tag{B.7}$$

Case 4

$$P^* \leq P_{t-1} \leq P_2, \quad \varepsilon_t = \varepsilon_2$$

which implies that $P^* \leq P_t \leq P_2$ and $I_{t-1} = I_t = 0$. Hence,

$$a\lambda_2(\varepsilon_2) = -b\bar{\lambda}_2 \tag{B.8}$$

$$a\mu_2(\varepsilon_2) = A + C - \varepsilon_2 - b\bar{\mu}_2 \tag{B.9}$$

Equations (B.1) - (B.9) can be solved for the eight unknowns, $\lambda_1(\varepsilon_t)$, $\lambda_2(\varepsilon_t)$, $\mu_1(\varepsilon_t)$, $\mu_2(\varepsilon_t)$, $\varepsilon_t = \{\varepsilon_1, \varepsilon_2\}$, as follows.

Solve (B.2) and (B.4) for $\lambda_1(\varepsilon_1)$ and $\lambda_1(\varepsilon_2)$ in terms of $\bar{\lambda}_1$. Then take expectations $\pi_1\lambda_1(\varepsilon_1) + \pi_2\lambda_1(\varepsilon_2) = \bar{\lambda}_1$, to get:

$$\bar{\lambda}_1 = [d(1 - \beta\bar{\lambda}_1) - b\bar{\lambda}_1][\frac{\pi_1}{a + d(1 - \beta\bar{\lambda}_1)} + \frac{\pi_2}{a}] \tag{B.10}$$

which is a quadratic equation in $\bar{\lambda}_1$. Letting $X = 1 - \beta\bar{\lambda}_1$, we may write (B.10) as:

$$f(X) = X^2[d + \pi_2 \frac{d}{a}(b + Bd)] + X[a + b - \pi_2 \frac{db}{a} - d(1 - \beta)]$$

$$- (a+b)$$

$$= 0 \tag{B.11}$$

Since $f(0) < 0$ and $f(1) > 0$ there exists a root X_1 such that $0 < X_1 < 1$. Hence, there exists a solution to (B.10) with $0 < \bar{\lambda}_1 < 1/\beta$. Now since $\bar{\lambda}_1 > 0$, either $\lambda_1(\varepsilon_1) > 0$ or $\lambda_1(\varepsilon_2) > 0$ (or both). Comparing (B.2) and (B.4) and noting that $\bar{\lambda}_2 \beta < 1$, we see that $\lambda_1(\varepsilon_1) > 0$ if and only if $\lambda_1(\varepsilon_2) > 0$. Hence,

$$0 < \lambda_1(\varepsilon_1) < 1$$

$$0 < \lambda_1(\varepsilon_1) < \bar{\lambda}_1 < \lambda_1(\varepsilon_2). \tag{B.12}$$

Proceeding in a similar way, we solve (B.3) and (B.5) for $\mu_1(\varepsilon_1)$ and $\mu_1(\varepsilon_2)$. By taking expectations, $\pi_1 \mu_1(\varepsilon_1) + \pi_2 \mu_1(\varepsilon_2) = \bar{\mu}_1$, we obtain:

$$\bar{\mu}_1 = \pi_1 \left[\frac{A + C - \varepsilon_1 - b\bar{\mu}_1}{a + d(1 - \beta\bar{\lambda})} \right] + \pi_2 \frac{[A + C - \varepsilon_2 - (b + \beta d)\bar{\mu}_1]}{a}. \tag{B.13}$$

By using the solution for $\bar{\lambda}_2$ obtained from (B.10) we can solve (B.13) for $\bar{\mu}_1$, after which we solve for $\mu_1(\varepsilon_1)$ and $\mu_1(\varepsilon_2)$ from (B.3) and (B.1). It can easily be shown that $\bar{\mu}_1 > 0$.

Following the same procedure, we can solve (3.18) and (3.20) for $\lambda_2(\varepsilon_1)$ and $\lambda_2(\varepsilon_2)$ to obtain $\lambda_2(\varepsilon_1) = \lambda_2(\varepsilon_2) = \bar{\lambda}_2 = 0$. Notice that this indicates that in the absence of inventories prices will be serially uncorrelated. In a similar way, we can solve (4.19) and (4.21) for $\mu_2(\varepsilon_1)$, $\mu_2(\varepsilon_2)$ and $\bar{\mu}_2 = \pi_1 \mu_2(\varepsilon_1) + \pi_2 \mu_2(\varepsilon_2)$.

Because the suggested solutions (4.11) and (4.12) should be continuous at P* for each realization of ε_t, we require that

$$\lambda_1(\varepsilon_1) P^* + \mu_1(\varepsilon_1) = \mu_2(\varepsilon_1)$$

$$\lambda_1(\varepsilon_2) P^* + \mu_1(\varepsilon_2) = \mu_2(\varepsilon_2) \tag{B.14}$$

or

$$P^* = \frac{\mu_2(\varepsilon_1) - \mu_1(\varepsilon_1)}{\lambda_1(\varepsilon_1)} = \frac{\mu_2(\varepsilon_2) - \mu_1(\varepsilon_2)}{\lambda_1(\varepsilon_2)}.$$

The latter equality may be verified directly by use of earlier expressions derived. Hence, P* is uniquely defined.

By multiplying the first equation in (B.14) by π_1 and the second equation in (4.26) by π_2 and adding we see that

$$P^* = \frac{\bar{\mu}_2 - \bar{\mu}_1}{\bar{\lambda}_1}$$

and by using (4.15) for $\mu_1(\varepsilon_1)$, (4.14) for $\lambda_1(\varepsilon_1)$ and (B.7) for $\mu_2(\varepsilon_1)$, we obtain:

$$P^* = \frac{b(\bar{\mu}_2 - \bar{\mu}_1) - \beta d \bar{\mu}_1}{\bar{\lambda}_1 (b + \beta d) - d} = \frac{\bar{\mu}_2 - \bar{\mu}_1}{\bar{\lambda}_1} \qquad (B.15)$$

Hence,

$$P^* = \frac{\beta \bar{\mu}_1}{1 - \beta \bar{\lambda}_1} > 0 \qquad (B.16)$$

since $\bar{\mu}_1 > 0$, which from (B.15) implies that $\bar{\mu}_2 > \bar{\mu}_1$.

We now verify that

$$P_1 \leq P_t \leq P^* \Rightarrow I_t \geq 0$$

$$P^* \leq P_t \leq P_2 \Rightarrow I_t = 0.$$

For $P_t \leq P^*$,

$$\beta E_t(P_{t+1}) - P_t = \beta[\bar{\lambda}_1 P_t + \bar{\mu}_1] - P_t = \beta \bar{\mu}_1 - P_t(1 - \beta \bar{\lambda}_1) \geq 0.$$

Hence, $I_t \geq 0$ for $P_t \leq P^*$.

For $P_t \geq P^*$,

$$\beta E_t(P_{t+1}) - P_t = \beta(\bar{\lambda}_2 P_t + \bar{\mu}_2) - P_t$$

$$= \beta\bar{\mu}_2 - P_t, \text{ because } \bar{\lambda}_2 = 0$$

$$\leq 0$$

since from (B.15) and (B.16) $P^* = \beta\bar{\mu}_2$. Hence $I(t) = 0$ for $P_t \geq P^*$.

References

Blinder, Alan S. (1982): "Inventories and Sticky Prices," *American Economic Review*, 72, 334-348.

Blinder, Alan S. and Stanley Fischer (1981): "Inventories, Rational Expectations and the Business Cycle," *Journal of Monetary Economics*, 8, 277-304.

Brennan, Michael J. (1958): "The Supply of Storage," *American Economic Review*, 48, 50-72.

Hodrick, R. J. and Edward C. Prescott, (1980): "Post-War U.S. Business Cycles: An Emprical Investigation," Manuscript, Carnegie-Mellon University.

Kohn, Meir (1978): "Competitive Speculation," *Econometrica*, 46, 1061-1076.

Lucas, Robert E. (1977): "Understanding Business Cycles," in *Stabilization of the Domestic and International Economy*, ed. by Karl Brunner and Alan H. Meltzer, New York: North-Holland.

Lucas, Robert E. and Edward C. Prescott (1971): "Investment Under Uncertainty," *Econometrica*, 39, 659-681.

Muth, John F. (1961): "Rational Expectations and the Theory of Price Movements," *Econometrica*, 29, 315-335.

Reagan, Patricia B. (1982): "Inventories and Price Behavior," *Review of Economic Studies*, 49, 137-142.

Reagan, Patricia. B. and Dennis P. Sheehan (1982): "The Stylized Facts About the Behavior of manufacturers' Inventories and Backorder Over the Business Cycle: 1959-1980," *Journal of Monetary Economics*, 9, 217-246.

Reagan, Patricia B. and Martin L. Weitzman (1982): "Price and Quantity Adjustment by the Competitive Industry," *Journal of Economic Theory*, 15, 410-420.

Sargent, Thomas J. (1979): "Tobin's q and the Rate of Investment in General Equilibrium," in *On the State of Macroeconomics*, ed. by Karl Brunner and Allan H. Meltzer, Amsterdam, North-Holland.

Sargent, Thomas J. (1979): *Macroeconomic Theory*, New York: Academic Press.

Scheinkman, J. A. and J. Schechtman (1982): "A Simple Competitive Model with Production and Storage," manuscript.

Turnovsky, Stephen J. (1978): "Stabilization Rules and the Benefits from Price Stabilization," *Journal of Public Economics*, 9, 37-57.

Wright, Brian D. and Jeffrey C. Williams (1982): "The Economic Role of Commodity Storage," *The Economic Journal*, 92, 596-614.

Zabel, Edward (1972): "Multiperiod Monopoly under Uncertainty," *Journal of Economic Theory*, 524-536.

CHAPTER III

TEMPORAL AGGREGATION AND THE
STOCK ADJUSTMENT MODEL OF INVENTORIES[1]

Lawrence J. Christiano
Federal Reserve Bank of Minneapolis and NBER

Martin Eichenbaum
Northwestern University and NBER

Abstract

This paper examines the quantitative importance of temporal aggregation bias in distorting parameter estimates and hypothesis tests in the production smoothing/buffer stock model of inventories. In particular, our results are consistent with the Mundlak-Zellner hypothesis that temporal aggregation bias can account for the slow speeds of adjustment typically obtained in such a model.

1. Introduction

Application of the stock adjustment model to the study of inventory behavior frequently produces implausibly low estimates of the speed of adjustment of actual to target inventories. For example, the parameter estimates reported by Feldstein and Auerbach (1976) imply that firms take almost 19 years to close 95 percent of the gap between actual and desired inventories. Application of the stock adjustment model to other problems such as the demand for money also yields implausibly low speeds of adjustment.

A variety of interesting explanations for these anomalous results exist. Blinder (1986), Eichenbaum (1984), and Maccini and Rossana (1984) explore different explanations for the slow estimated speed of

---

[1]This is a revised version of Section 3 of our paper, "Temporal Aggregation and Structural Inference in Macroeconomics," in <u>Bubbles and Other Essays</u>, ed. by Brunner and Meltzer, Carnegie-Rochester Conference Series on Public Policy, vol. 26, Spring 1987. We are grateful to Tryphon Kollintzas and Kenneth D. West for comments. The views expressed herein are those of the authors and not necessarily those of the Federal Reserve Bank of Minneapolis or the Federal Reserve System.

adjustment of inventories. Goodfriend (1985) discusses this problem with respect to the demand for money. In this paper we explore the possibility that estimated slow speeds of adjustment reflect temporal aggregation bias. Mundlak (1961) and Zellner (1968) show theoretically that, if agents make decisions at intervals of time that are finer than the data sampling interval, then the econometrician could be led to underestimate speeds of adjustment. This is consistent with findings reported by Bryan (1967), who applies the stock adjustment model to bank demand for excess reserves. Bryan finds that when the model is applied to weekly data, the estimated time to close 95 percent of the gap between desired and actual excess reserves is 5.2 weeks. When the model is applied to monthly aggregated data, the 95 percent closure time is estimated to be 28.7 months.

Our strategy for investigating the role played by time aggregation on the empirical estimate of speed of adjustment is as follows. First, we construct a continuous time equilibrium rational expectations model of inventories and sales. The model rationalizes a continuous time inventory stock adjustment equation. Using techniques developed by Hansen and Sargent (1980a, 1981), we estimate the model using monthly data on inventories and sales in the nondurable manufacturing sector. The parameter estimates from the continuous time model imply that firms close 95 percent of the gap between actual and desired inventories in 17 days. We then estimate an analogous discrete time model using monthly, quarterly and annual data. The parameter estimates obtained using monthly data imply that it takes firms 46 days to close 95 percent of the gap between actual and desired inventories. The analogous figure obtained with annual data implies that it takes firms 1,980 days to close 95 percent of the gap between actual and desired inventories. These results indicate that estimates of speed of adjustment are very sensitive to the effects of time aggregation.

Unfortunately, we cannot claim that temporal aggregation effects account for the statistical shortcomings of existing stock adjustment models. Both the discrete and continuous time versions of our equilibrium stock adjustment model impose strong overidentifying restrictions on the data. Using a variety of tests and diagnostic devices, we find substantial evidence against these restrictions. In addition, we find no evidence that the overall fit for the continuous time model is superior to that of the discrete time model.

In Section 2 we formulate a continuous time equilibrium model of employment, inventories of finished goods and output. In Section 3 we discuss an estimation strategy which explicitly takes the temporal aggregation problem into account. Section 4 presents the empirical results.

2. <u>A Continuous Time Model of Inventories, Output and Sales</u>

In this section we discuss a modified continuous time version of the model in Eichenbaum (1984). Our model is designed to nest, as a special case, the model considered by Blinder (1981, 1986) and Blinder and Holtz-Eakin (1984). We take the model to be representative of an interesting class of inventory models. An important virtue of our model is that it provides an explicit equilibrium rationale for a continuous time version of the stock adjustment equation for inventories. An additional advantage of proceeding in terms of an equilibrium model is that we are able to make clear both the theoretical underpinnings and the weaknesses of an important class of inventory models which has appeared in the literature.

Consider a competitive representative household that ranks alternative streams of consumption and leisure using this utility function:[2]

$$E_t \int_0^\infty e^{-r\tau} \{u(t+\tau)s(t+\tau) - .5A[s(t+\tau)]^2 - N(t+\tau)\} d\tau. \qquad (1)$$

---

[2]The fact that we specify utility to be linear in leisure warrants some discussion because it appears to be inconsistent with findings in two recent studies. Our specification implies that leisure in from the point of view of the representative consumer, leisure in any period is a perfect substitute for leisure in any other period. MaCurdy (1981) and Altonji (1986) argue, on the basis of panel data, that, from the point of view of private agents, leisure in any period is an imperfect substitute for leisure in any other period. Rogerson (1984) and Hansen (1985) describe conditions under which the assumption that the representative consumer's utility function is linear in leisure is consistent with any degree of intertemporal substitutability at the level of private agents.

In (1),

- t = the time unit, measured in months,
- E_t = the linear least squares projection operator, conditional on the time t information set,
- $s(t)$ = time t consumption of the single nondurable consumption good,
- $N(t)$ = total work effort at time t,
- $u(t)$ = a stochastic disturbance to the marginal utility of consumption at time t, and
- A, r = positive constants.

We now specify the technology for producing new consumption goods and storing inventories of finished goods. Let $Q(t)$ denote the total output of new consumption goods at time t. The production function for $Q(t)$ is defined implicitly by

$$N(t) = a/2 \, Q(t)^2 + C_I(t), \tag{2}$$

where a is a positive scalar and $C_I(t)$ represents inventory holding costs. In order to accommodate two different types of costs associated with inventories that have been considered in the literature, we suppose that total inventory costs, measured in units of labor, are given by

$$C_I(t) = (b/2)[s^*(t) - cI(t)]^2 + v(t)I(t) + (e/2)I(t)^2, \tag{3}$$

where b, c, and e are positive scalars, $v(t)$ is a stochastic shock to marginal inventory holding costs, $s^*(t)$ denotes time t sales of the good and $I(t)$ is the stock of inventories at date t. The last two terms in (3) correspond to the inventory holding cost function adopted by Blinder (1981, 1986) and Blinder and Holtz-Eakin (1984), among others. This component of costs reflects the physical costs of storing inventories of finished goods. The first term in (3) reflects the idea that there are costs, denominated in units of labor, associated with allowing inventories to deviate from some fixed proportion of sales. Blanchard (1983, p. 378) provides an extensive motivation of this component of inventory costs. Similar cost functions appear in Eichenbaum (1984), McCallum (1984) and Eckstein and Eichenbaum (1985).

The link between current production, inventories of finished goods and sales is given by

$$Q(t) = s*(t) + DI(t), \qquad (4)$$

where D is the derivative operator $[Dx(t) = dx(t)/dt]$.

It is well known that, in the absence of externalities or similar types of distortions, rational expectations competitive equilibria are Pareto optimal. Since our representative consumer economy has a unique Pareto optimal allocation, we could solve directly for the competitive equilibrium by considering the relevant social planning problem (see Lucas and Prescott (1971), Hansen and Sargent (1980b) and Eichenbaum, Hansen and Richard (1985)). On the other hand, there are a variety of market structures which will support the Pareto optimal allocation. In the interest of preserving comparability with other papers in the inventory literature, we find it convenient to work with a particularly simple market structure that supports this allocation. As in Sargent (1979), we require only competitive spot markets for labor and the consumption good to support the Pareto optimal allocation.[3]

Suppose that the representative consumer chooses contingency plans for $s(t+\tau)$, $\tau \geq 0$, to maximize (1) subject to the sequence of budget constraints

$$P(t+\tau)s(t+\tau) = N(t+\tau) + \pi(t+\tau). \qquad (5)$$

---

[3] It is of interest to contrast our model with the equilibrium model in Sargent (1979, Chapter XV). In that model, the representative agent's utility function is linear in consumption and quadratic in leisure. As a result, the interest rate on risk-free securities, denominated in units of the consumption good, is constant. In our model, the representative agent's utility function is quadratic in consumption, with the result that the interest rate on risk-free securities, denominated in units of the consumption good, is time varying and stochastic. This feature of our model is attractive in view of the apparent nonconstancy of real interest rates in the U.S. In order to remain within the linear-quadratic framework, we specify utility to be linear in leisure. This implies that the interest rate on risk-free securities, denominated in units of leisure, is constant.

In (5), $P(t)$ = the price of the consumption good, denominated in labor units, and $\pi(t)$ = lump sum dividend earnings of the household, denominated in labor units.

Solving the representative consumer's problem, we obtain the following inverse demand function:

$$P(t) = -As(t) + u(t). \tag{6}$$

Given the very simple structure of relation (6), it is important to contrast our specification of the demand function with different specifications that have been adopted in the literature. In constructing empirical stock adjustment models, most analysts abstract from modelling demand. Instead, the analysis is conducted assuming a particular time series representation for an exogenous sales process (see, for example, Feldstein and Auerbach (1976) or Blanchard (1983)). Our model is consistent with this practice when A is very large. To see this, rewrite (6) as

$$s(t) = -(1/A)P(t) + \eta(t), \tag{6'}$$

where $\eta(t) = -(1/A)u(t)$. The assumptions we place on $u(t)$ below guarantee that $\eta(t)$ has a time series representation of the form $\gamma(D)\eta(t) = \nu(t)$, where $\nu(t)$ is continuous time white noise, uncorrelated with past values of $s(t)$ and $I(t)$. Also, $\gamma(t)$ is a finite-ordered polynomial satisfying the root condition required for covariance stationarity. If A is very large (infinite), then sales have the reduced form time series representation $\gamma(D)s(t) = \nu(t)$. This is the continuous time analogue of the assumption, made in many stock adjustment models, that sales are exogenous stochastic processes in the sense of not being Granger-caused by the actions of the group of agents who make inventory decisions. (Our empirical results indicate that the assumption of one-way Granger-causality from sales to inventory stocks is reasonably consistent with the data.)

Other authors, like Blinder (1986) and Eichenbaum (1984), begin their analysis by postulating the industry demand curve (6). Our analysis provides an equilibrium interpretation of this demand specification. In so doing, we are forced to confront the strong assumptions implicit in (6). For example, we implement our model on nondurable manufacturing shipment and inventory data. This choice of data was dictated by the desire for our results to be comparable with those appearing in the relevant literature. Notice, however, that manufacturers' shipments do not enter directly as arguments into consumers' utility functions. Rather, they represent sales from manufacturers to wholesalers and retailers, who in turn sell them to households. Consequently, objective function (1) consolidates the wholesale, retail and household sectors. We know of no empirical justification for this assumption. By focusing on nondurable manufacturers, we place more faith than we care to on the stability of their relation to wholesalers and retailers. For example, shifts through time in the pattern of inventory holdings between manufacturers and retailers and wholesalers would have effects on our empirical results that are hard to predict. At the same time, they do not represent phenomena that we wish to model in this paper.[4] (In future research, we plan to avoid this type of problem by consolidating data from the wholesale, retail and manufacturing sectors.)

We assume that the representative firm seeks to maximize its expected real present value. The firm distributes all profits in the form of lump sum dividends to consumers. The firm's time t profits are equal to

$$\pi(t) = P(t)s^*(t) - N(t). \qquad (7)$$

Substituting (2), (3) and (4) into (7), we obtain

$$\pi(t) = P(t)s^*(t) - (a/2)[s^*(t)+DI(t)]^2 - (b/2)[s^*(t)-cI(t)]^2 \\ - v(t)I(t) - (e/2)I(t)^2. \qquad (8)$$

The firm chooses contingency plans for $s^*(t+\tau)$ and $DI(t+\tau)$, $\tau \geq 0$, to maximize

---

[4]See Dimelis and Kollintzas (1988) for a similar model that distinguishes between raw material and final product good inventories.

$$E_t \int_0^\infty e^{-r\tau} \pi(t+\tau) d\tau \qquad (9)$$

given $I(t)$, the laws of motion of $v(t)$ and $u(t)$, beliefs about the law of motion of industry wide sales, $s^*(t)$.[5] In a rational expectations equilibrium, these beliefs are self-fulfilling. Sargent (1979, p. 375) describes a simple procedure for finding rational expectations equilibria in linear-quadratic, discrete time models. The discussion in Hansen and Sargent (1980a) shows how to modify Sargent's solution procedure to accommodate our continuous time setup. Briefly, the procedure is as follows.[6] Write,

$$F[I(t), DI(t), s^*(t), v(t), P(t), t] = e^{-rt} \pi(t), \qquad (10)$$

so that (9) can be written as

$$E_t \int_0^\infty F[I(t+\tau), DI(t+\tau), s^*(t+\tau), v(t+\tau), P(t+\tau), \tau] d\tau, \qquad (11)$$

by choice of $DI(t+\tau)$, $s^*(t+\tau)$, $\tau \geq 0$, subject to $I(t)$ and the laws of motion of $v(t)$ and $P(t)$. Notice that the principle of Certainty Equivalence applies to this problem. Accordingly, we first solve a version of (11) in which future random variables are equated to their time t conditional expectation. Then we use a continuous time version of the Weiner-Kolmogorov forecasting formula to express the time t conditional expectation of time $t + \tau$ variables in terms of elements of agents' time t information set.

---

[5] To avoid proliferating notation, we do not formally distinguish between variables chosen by individual households and firms and their economy wide counterparts. Nevertheless, the distinction between them plays an important role in the model. By assumption, agents are perfectly competitive and view economy wide variables, such as $P(t)$ and economy wide sales and inventories, parametrically.

[6] See Hansen and Sargent (1980a), which shows that this procedure yields the unique optimal solution to (9).

The variational methods discussed by Luenberger (1969) imply that the firm's Euler equations for $s(t)$ and $I(t)$ are

$$\partial F/\partial s^*(t) = 0 \qquad (12a)$$

and

$$\partial F/\partial I(t) = D\{\partial F/\partial DI(t)\}. \qquad (12b)$$

These imply, respectively

$$P(t) - (a+b)s^*(t) - aDI(t) + bcI(t) = 0 \qquad (13a)$$

and

$$aD^2I(t) - raDI(t) - (c^2b+e)I(t) + aDs^*(t) + (cb-ra)s^*(t) = v(t). \quad (13b)$$

In a rational expectations competitive equilibrium, $P(t)$ must satisfy (6), with $s(t) = s^*(t)$. Substituting (6) into (13a) and replacing $s^*(t)$ by $s(t)$, we obtain

$$s(t) = -[a/(a+b+A)]DI(t) + [bc/(a+b+A)]I(t) + [1/(a+b+A)]u(t). \quad (14)$$

It is convenient to collapse (13b) and (14) into one differential equation in $I(t)$. Substituting $s(t)$ and $Ds(t)$ from (14) into (13b), we obtain

$$(D-\lambda)[D-(r-\lambda)]I(t) = \frac{(a+b+A)}{a(b+A)} v(t) - \frac{1}{(b+A)}[(bc-ra)/a+D]u(t), \quad (15a)$$

where

$$\lambda = .5r + (k+.25r^2)^{1/2} \qquad (15b)$$

and

$$k = [(a+b+A)/a(b+A)]\{[bc[c(a+A)+ra]/(a+b+A)]+e\}. \qquad (15c)$$

Since $k > 0$, it follows from (15b) that $\lambda > 0$ is real. Moreover, it is easy to verify that $r - \lambda = .5r - [k + .25r^2]^{1/2}$.[7] Solving the stable root $(r-\lambda)$ backward and the unstable root λ forward in (15a), we obtain

$$DI(t) = (r-\lambda)I(t) - \frac{a+b+A}{a(b+A)} \int_0^\infty e^{-\lambda\tau} v(t+\tau)d\tau$$

$$+ \frac{1}{b+A} \int_0^\infty e^{-\lambda\tau} [(cb-ra)/a+D]u(t+\tau)d\tau$$

$$= (r-\lambda)I(t) - \frac{a+b+A}{a(b+A)} \int_0^\infty e^{-\lambda\tau} v(t+\tau)d\tau - \frac{1}{b+a} u(t)$$

$$+ \frac{1}{b+a} [\frac{bc}{a} - (r-\lambda)] \int_0^\infty e^{-\lambda\tau} u(t+\tau)d\tau, \qquad (16)$$

where the second equality is obtained using integration by parts. Substituting (16) into (14), we obtain

$$s(t) = \frac{bc - a(r-\lambda)}{a+b+A} I(t) + \frac{1}{b+A} \int_0^\infty e^{-\lambda\tau} v(t+\tau)d\tau + \frac{1}{b+A} u(t)$$

$$- \frac{1}{(b+A)(a+b+A)} [(bc/a) - (r-\lambda)] \int_0^\infty e^{-\lambda\tau} u(t+\tau)d\tau. \qquad (17)$$

Equations (16) and (17) are the equilibrium laws of motion for inventory investment and consumption in the perfect foresight version of our model. Before allowing for uncertainty, we discuss some qualitative features of this equilibrium.

First, suppose that the parameter b is equal to zero and there are no technology shocks. This is the model considered by Blinder (1981, 1986) and Blinder and Holtz-Eakin (1984). The role of inventories in this version of the model is to smooth production in the sense that inventory investment is negatively related to current demand shocks and positively related to expected future demand shocks (see (16) and recall that $r-\lambda < 0$). As Blinder (1986) points out, production

---

[7] To see that $\lambda > 0$ consider $f(k) = .5r - [k + .25r^2]^{1/2}$ and note that $f(0) = 0$ and $f'(k) < 0$ for $k \geq 0$.

smoothing, as defined here, does not necessarily imply that the variance of sales will exceed that of production. For example, if the serial correlation structure of u(t) were such that a jump in u(t) typically implies a large increase in u(t) in the future, then the current jump in u(t) could lead to an increase in inventory investment, as well as sales. We rule out these types of u(t) processes below. Consequently, production smoothing in our model implies that the variance of production is lower than the variance of sales when b = v(t) = 0.

Second, suppose that there are no preference shocks. Then the role of inventories is to smooth sales. To see this, notice that inventory investment depends negatively on current and future shocks to the inventory holding cost function. The firm holds less inventories when the marginal cost of holding inventories increases. Suppose that inventory holding costs are viewed as general shocks to production costs. Firms will use inventories to smooth production costs, as opposed to production levels, over time in the face of stable demand for their product. For the kinds of production cost shocks that we consider in this paper, this implies that the variance of sales will be smaller than the variance of production.

A slightly different way of seeing these points is to remember that the competitive equilibrium solves the problem of a fictitious social planner/representative consumer. The representative consumer has a utility function which is locally concave in consumption so that, other things equal, a smooth consumption path is preferred. If preference shocks predominate, we would expect sales/consumption to be volatile relative to production. On the other hand, if technology shocks predominate, we would expect sales/consumption to be smooth relative to production. Blinder (1981, 1986) and West (1986) document the fact that, at least for post-World War II data, the variance of production exceeds the variance of sales/consumption. This suggests that the primary role of inventories is to smooth sales rather than production levels.

We now consider the equilibrium of the system in the uncertainty case. In order to derive explicit expressions for the equilibrium laws of motion of the system, we parameterize the stochastic laws of motion of the shocks to preferences and technology. To this end, we assume that $u(t)$ and $v(t)$ have the joint AR(1) structure

$$u(t) = \epsilon_1(t)/(\beta+D) = \int_0^\infty e^{-\beta\tau}\epsilon_1(t-\tau)d\tau \tag{18a}$$

and

$$v(t) = \epsilon_2(t)/(\alpha+D) = \int_0^\infty e^{-\alpha\tau}\epsilon_2(t-\tau)d\tau, \tag{18b}$$

where α and β are positive scalars. The vector $\epsilon(t) = [\epsilon_1(t)\epsilon_2(t)]'$ is the continuous time linear least squares innovation in $[u(t)v(t)]'$, $E\epsilon(t)\epsilon(t-\tau)' = \delta(\tau)\tilde{V}$ is a positive definite 2×2 symmetric matrix and $\delta(\tau)$ is the Dirac delta generalized function.

Given the above specification for the shocks, it is obvious that, for $\tau \geq 0$,

$$E_t u(t+\tau) = \int_0^\infty e^{-\beta s}\epsilon_1(t+\tau+s)ds = e^{-\beta\tau}\int_0^\infty e^{-\beta s}\epsilon_1(t-s)ds = e^{-\beta\tau}u(t). \tag{19a}$$

Similarly,

$$E_t v(t+\tau) = e^{-\alpha\tau}v(t). \tag{19b}$$

Simple substitution from (19) yields

$$E_t \int_0^\infty e^{-\lambda\tau}u(t+\tau)d\tau = u(t)/(\beta+\lambda)$$

and

$$E_t \int_0^\infty e^{-\lambda\tau}v(t+\tau)d\tau = v(t)/(\alpha+\lambda).$$

Substituting these expressions into (16) and (17), we obtain the equilibrium laws of motion for $DI(t)$ and $s(t)$,

$$DI(t) = (r-\lambda)I(t) - \frac{a+b+A}{a(b+A)(\alpha+\lambda)} v(t) + \frac{(cb-ra) - a\beta}{a(b+A)(\beta+\lambda)} u(t) \qquad (20a)$$

$$s(t) = \frac{bc - a(r-\lambda)}{a+b+A} I(t) + \frac{v(t)}{(b+A)(\alpha+\lambda)} + \frac{a}{(b+A)(a+b+A)} \frac{[(bc-ra)-\beta a]}{a(\beta+\lambda)}$$

$$+ [\frac{1}{a+b+A}]u(t). \qquad (20b)$$

It is convenient to write the equilibrium laws of motion for $I(t)$ and $s(t)$ in the form of a continuous time moving average of $\epsilon_1(t)$ and $\epsilon_2(t)$. Substituting (18) into (20) and rearranging, we obtain, in operator notation,

$$\begin{bmatrix} I(t) \\ s(t) \end{bmatrix} = \theta(D)^{-1}\tilde{C}(D)\epsilon(t), \qquad (21')$$

where

$$\theta(D) = (\alpha+D)(\beta+D)[D-(r-\lambda)] \qquad (22)$$

$$\tilde{C}(D) = \tilde{C}_0 + \tilde{C}_1 D + \tilde{C}_2 D^2 \qquad (23)$$

$$\tilde{C}_0 = \begin{bmatrix} q_1\alpha & q_2\beta \\ \frac{\alpha[q_1 bc-(r-\lambda)]}{a+b+A} & \frac{q_2 bc\beta}{a+b+A} \end{bmatrix}$$

$$\tilde{C}_1 = \begin{bmatrix} q_1 & q_2 \\ \frac{-aq_1(\alpha-\frac{bc}{a})+\alpha-(r-\lambda)}{a+b+A} & \frac{-aq_2(\beta-bc/a)}{a+b+A} \end{bmatrix}$$

$$\tilde{C}_2 = \begin{bmatrix} 0 & 0 \\ \frac{1-aq_1}{a+b+A} & \frac{-aq_2}{a+b+A} \end{bmatrix}$$

$$q_1 = \frac{(cb-ra)-a\beta}{a(\lambda+\beta)(b+A)} \tag{24}$$

and

$$q_2 = \frac{-(a+b+A)}{a(b+A)(\lambda+\alpha)}.$$

We find it useful to write (21)' as

$$\begin{bmatrix} I(t) \\ s(t) \end{bmatrix} = \theta(D)^{-1} C(D) e(t), \tag{21}$$

where $e(t) = \tilde{C}_0 \epsilon(t)$, $C(D) = \tilde{C}(D)\tilde{C}_0^{-1}$, and $Ee(t)e(t)' = \delta(\tau)V = \delta(t)\tilde{C}_0 \tilde{V} \tilde{C}_0'$. With this definition of $C(D)$ and $e(t)$, equations (21)-(24) summarize all of the restrictions that our model imposes on the continuous time Wold MA representation of $I(t)$ and $s(t)$.

We conclude this section by showing that our model is consistent with a stock adjustment equation for inventories. Let $I(t)^*$ denote the aggregate level of inventories such that if $I(t) = I(t)^*$, then actual inventory investment, $DI(t)$, is equal to zero. $I(t)^*$ is taken to be the level of desired or target, inventories. Relation (20a) implies that

$$I(t)^* = \frac{a+b+A}{(r-\lambda)a(b+A)(\alpha+\lambda)} v(t) - \frac{(bc-ra)-a\beta}{a(r-\lambda)(b+A)(\beta+\lambda)} u(t). \tag{25}$$

Substituting (25) into (20a), we obtain a stock adjustment equation for inventory investment,

$$DI(t) = (\lambda-r)[I(t)^* - I(t)]. \tag{26}$$

We require a measure of the speed of adjustment which can be compared with similar measures reported in the literature. In order to make this concept precise, we imagine, counterfactually, that movements in $I(t)^*$ can be ignored over an interval $\tau \in (t,t+1)$, so that $I(\tau)^* = I(t)^*$ for $\tau \in (t,t+1)$. Then the solution to (26) is

$$I(t+\tau) - I(t)^* = e^{-(\lambda-r)\tau}[I(t)-I(t)^*]. \qquad (27)$$

Relation (27) gives rise to an interesting summary statistic regarding the speed of adjustment of actual to target inventories. In particular, the number of days required to close 95 percent of the gap between actual and target inventories is

$$T^c = -30 \ [\log \ (1-.95)]/(\lambda-r), \qquad (28)$$

where 30 is approximately the number of days in a month.

Given the estimates of the structural parameters, it is straightforward to calculate this statistic. In the next section we discuss a strategy for estimating the parameters of our model from discrete data. In addition, we formulate a discrete time version of the model which is useful for estimating speeds of adjustment under the assumption that agents' decision intervals coincide with the data sampling interval.

3. Estimation Issues

In this section, we discuss a strategy for estimating the continuous time model of Section 2 from discrete observations on inventories and sales. Since our estimator corresponds to the one discussed in Hansen and Sargent (1980a), we refer the reader to that paper for technical details. Christiano and Eichenbaum (1985) provide additional details for the model considered here. In this section, we also display a discrete time version of our basic model and describe a method for estimating its parameters. By estimating both models, we are able to derive an empirical measure of the effects of temporal aggregation on speed of adjustment estimates.

We now describe the procedure used to estimate the parameters of the continuous time model described in Section 2. This procedure takes into account the fact that the inventory data are point-in-time and measured at the beginning of the sampling interval, while sales are averages over the month.

Our estimation strategy involves maximizing an approximation of the Gaussian likelihood function of the data with respect to the unknown parameters, ς, which we list explicitly in Section 4. The approximation we use is the frequency domain approximation studied extensively in Hannan (1970). Hansen and Sargent (1980a) show how to use this approximation to estimate continuous time linear rational expectations models from discrete data records.

One way to describe our estimation strategy exploits the observation that estimation of a continuous time model actually is a special case of estimating a constrained discrete time model. Recall from the discussion of Section 2 that ς implies a continuous time ARMA model, characterized by the polynomials $\theta(D)$ and $C(D)$ and a symmetric matrix, V (see (21)-(24)). This continuous time series representation implies a particular discrete time series representation for the sampled, averaged data. In Christiano and Eichenbaum (1987, Theorem 2) we characterized this discrete time representation by a scalar third-order polynomial $\theta^c(L)$, a third-order 2×2 matrix polynomial $\bar{C}^c(L)$ and an innovation variance matrix V^c. Here, L is the lag operator where $L^j x(t) = x(t-j)$. The polynomial θ^c satisfies $\theta^c(e^{-\alpha}) = \theta^c(e^{-\beta}) = \theta^c(e^{r-\lambda}) = 0$ and $\theta^c(0) = 1$. Also, $\det \bar{C}^c(z) = 0$ implies $|z| \geq 1$. If we let $Y(t)$ denote the measured data on inventories and sales ($Y(t)$ is defined precisely below), then the time series representation of ($Y(t)$, integer t) is

$$[1+\theta_1^c L+\theta_2^c L^2+\theta_3^c L^3]Y(t) = [I+\bar{C}_1^c L+\bar{C}_2^c L^2+\bar{C}_3^c L^3]u_t,$$

where u_t the white noise innovation in $Y(t)$ with variance V^c.

Given θ^c, $\bar{C}^c$ and V^c it is possible to compute the spectral density of the data, $S_y(z;\varsigma)$, which is one of the two ingredients of the spectral approximation to the likelihood function. It can be shown that $S_y(z;\varsigma)$ is given by

$$S_y(z;\varsigma) = \bar{C}^c(z)V^c\bar{C}^c(z^{-1})'/\theta^c(z)\theta^c(z^{-1}),$$

for $z = e^{-iw}$, $w \in (-\pi,\pi)$.

The other ingredient of the spectral approximation to the likelihood function is the periodogram of the data. We denote the available data by $\{Y(T),\ t=1,2,\ldots,T\}$. Here, $Y(t) = [I(t), \bar{s}(t)]'$, where $\bar{s}(t)$ denotes average sales:

$$\bar{s}(t) = \int_0^1 s(t+\tau)d\tau. \qquad (29)$$

The periodogram of the data at frequency w_j, $I(w_j)$, is

$$I(w_j) = (1/T)Y(w_j)Y(w_j)^H,$$

where H denotes the Hermetian transpose and

$$Y(w_j) = \sum_{t=1}^T Y(t)e^{-iw_j t}.$$

Here $w_j = 2\pi j/T$, $j = 1, 2, \ldots, T$. Given these expressions for $S_y(z;\varsigma)$ and $I(w_j)$, we can compute the spectral approximation to the likelihood function,

$$L_T(\varsigma) = -T \log 2\pi - .5 \sum_{j=1}^T \log \det [S(e^{-iw}j;\varsigma)]$$

$$- .5 \sum_{j=1}^T \text{trace } [S(e^{-iw}j;\varsigma)^{-1} I(w_j)]. \qquad (30)$$

Since the likelihood function (30) is a known function of the data and the parameters of the model, it can be maximized with respect to those parameters. We obtain an estimate of the variance-covariance matrix of the estimated coefficients by computing the negative of the inverse of the second derivative of L_T with respect to ς, evaluated at the estimated values of ς.

We now consider the problem of estimating a discrete time version of the model. Accordingly, we suppose that the representative consumer maximizes

$$E_t \sum_{j=0}^{\infty} \phi^j \{u(t+j)s(t+j) - .5As(t+j)^2 - N(t+j)\}, \tag{31}$$

subject to (5) by choice of linear contingency plans for s(t) and N(t). The parameter ϕ is a subjective discount factor that is between zero and one. As before, the solution to the consumer's problem is given by the inverse demand function (6).

The representative competitive firm chooses linear contingency plans for s*(t) and I(t) to maximize

$$E_t \sum_{j=0}^{\infty} \phi^j \{P(t+j)s^*(t+j) - (a/2)[s^*(t+j) + I(t+j) - I(t+j-1)]^2$$

$$- (b/2)[s^*(t+j) - cI(t+j)]^2 - v(t+j)I(t+j) - (e/2)I(t+j)^2\}, \tag{32}$$

subject to I(t) given and the laws of motion of v(t) and P(t). We suppose that the shocks to technology and preferences have a discrete time AR(1) representation:

$$u(t) = \mu u(t-1) + \epsilon_1(t) \tag{33a}$$

and

$$v(t) = \rho v(t-1) + \epsilon_2(t), \tag{33b}$$

where $|\mu| < 1$ and $|\rho| < 1$. Also, $\epsilon(t) = (\epsilon_1(t), \epsilon_2(t))'$ is a vector white noise that satisfies

$$E\epsilon(t)\epsilon(t-\tau)' = \begin{cases} \Omega, & \tau \text{ equal to zero,} \\ 0, & \tau \text{ otherwise.} \end{cases} \tag{34}$$

The model summarized by (31)-(34) is the discrete time version of our continuous time model in that, essentially, it has been obtained by replacing the D operator by its approximation, 1 - L. An alternative would have been to specify the discrete time model so that the implied reduced form time series representation for inventories and sales would be an ARMA of the same order as that predicted by the continuous time model. In order to do this, we would have to abandon the assumption that u(t) and v(t) have first-order autoregressive representations or change other basic features of the discrete time model. This is an important point which we will return to in Section 4.

Christiano and Eichenbaum (1985) show that the equilibrium laws of motion for inventories and sales are given by

$$I(t) = \Psi I(t-1) + hu(t) + gv(t) \tag{35a}$$

$$s(t) = -(a-bc)/(a+b+A)I(t) + a/(a+b+A)I(t-1) + [1/(a+b+A)]u(t), \tag{35b}$$

where

$$h = \frac{-1}{a[b(c+1)+A]}\{\Psi(a-bc)+[(a-bc)\Psi-a]\Psi\phi\mu/(1-\Psi\phi\mu)\}, \tag{35c}$$

$$g = \frac{-(a+b+A)\Psi}{a[b(c+1)+A](1-\Psi\phi\rho)},$$

$$\Psi + 1/(\Psi\phi) = \frac{-(a+b+A)}{\phi a[b(c+1)+A]} \left[\frac{\phi a^2 + (a-bc)^2}{a+b+A} - (a+bc^2+e+\phi a) \right],$$

and $|\Psi| < 1$.

The relevant measure of the speed of adjustment of inventories which can be compared to the measure which emerges from the continuous time model is

$$T^d = X[\log(.05)]/\log \Psi, \tag{28'}$$

where X is the number of days in the data sampling interval.

It is convenient to write the equilibrium law of motion for s(t) and I(t) in the form of a moving average representation of the discrete time innovations to agents' information sets. Substituting (33) into (35) and rearranging, we obtain

$$\begin{bmatrix} I(t) \\ s(t) \end{bmatrix} = \theta^d(L)^{-1} \bar{C}^d(L) \epsilon(t), \tag{36}$$

where

$$\theta^d(L) = (1-\rho)(1-\mu L)(1-\Psi L) \tag{37}$$

$$\bar{C}^d(L) = \bar{C}_0^d + \bar{C}_1^d L + \bar{C}_2^d L^2$$

$$\bar{C}_0^d = \begin{bmatrix} h & g \\ \frac{1-(a-bc)h}{a+b+A} & \frac{g(bc-a)}{a+b+A} \end{bmatrix} \tag{38}$$

$$\bar{C}_1^d = \begin{bmatrix} -h\rho & -g\mu \\ \frac{(ah-\Psi)-\rho[1-(a-bc)h]}{a+b+A} & \frac{g[a-\mu(bc-a)]}{a+b+A} \end{bmatrix}$$

$$\bar{C}_2^d = \begin{bmatrix} 0 & 0 \\ \frac{-\rho(ah-\Psi)}{a+b+A} & \frac{-g\mu a}{a+b+A} \end{bmatrix}.$$

Given these relations, the free parameters of the discrete time model can be estimated by maximizing Hannan's spectral approximation to the likelihood function.

We are now in a position to demonstrate some of the possible sources of temporal aggregation bias in estimates of the speed of adjustment. Relations (21)-(24) and (36)-(38) summarize the restrictions on the continuous and discrete time Wold representation imposed by the continuous and discrete versions of the model, respectively. It can be shown that the continuous and discrete time models imply that $I(t)$ and $s(t)$ have continuous and discrete time VAR(2) representations, respectively. For example, to see this for the continuous time model, notice that (21)-(24) imply

$$\det C(D) = (\alpha+D)(\beta+D)[D-(r-\lambda)]/(\lambda-r)a(b+A). \tag{39}$$

Premultiplying (21) by $C(D)^{-1} = C(D)^a / \det C(D)$, we obtain

$$(\lambda-\alpha)a(b+A)C(D)^a Y(t) = e(t). \qquad (40)$$

Here $C(D)^a$ denotes the adjoint matrix of $C(D)$. Thus $\{Y(t)\}$ is a pure VAR(2) in continuous time. However, Theorem 1 of Christiano and Eichenbaum (1987) implies that sampled and averaged $\{Y(t)\}$ is a discrete time ARMA(2,2) process. One moving average term is due to sampling and the other is due to averaging. We choose not to focus upon this representation of the discrete data because its AR part requires stronger than usual restrictions to ensure identification (see Christiano and Eichenbaum (1985), pp. 29-31). Instead, we focus on an alternative reduced form representation for the data which emerges from the continuous time model,

$$\theta^c(L)Y(t) = [I + C_1^c L + C_2^c L^2 + C_3^c L^3] e^c(t), \qquad (41)$$

where $e^c(t)$ is the innovation in $Y(t)$ which has covariance matrix V^c. Here $\det C^c(L) = \theta^c(L)\kappa(L)$, where $\kappa(L)$ is a second-order polynomial in the lag operator L. The presence of $\kappa(L)$ is a symptom of the effects of sampling and of averaging s(t). Since $\det C^c(L)$ is not proportional to $\theta^c(L)$, the sampled representation is not VAR(2). As we indicated, it is vector ARMA (2,2). Christiano and Eichenbaum (1985) discuss the mapping between the representations (40) and (41).

Of course, the discrete time model remains a VAR(2). It is useful to write the reduced form of the discrete model in a manner that is analogous to (41). Define $e^d(t) = \bar{C}_0 \epsilon(t)$ and $C^d(L) = \bar{C}^d(L)(\bar{C}_0^d)^{-1}$. Then (36) implies that the reduced form representation for Y(T) emerging from the discrete time model is

$$\theta^d(L)Y(t) = [I + C_1^d L + C_2^d L^2] e^d(t), \qquad (42)$$

where the first row of C_2^d is composed of zeros. We denote the covariance matrix of $e^d(t)$ by V^d.

Comparing (41) and (42), we see that the moving average component of the reduced form for the discrete model is of smaller order than that of the continuous time model. Again, this reflects the fact that the continuous time and discrete time models have <u>different</u> implications for measured data. Not surprisingly, estimation of the two models will yield different estimates of the underlying structural parameters and speeds of adjustment of actual to target inventories.

4. Empirical Results

In this subsection, we report empirical results obtained from estimating four different models. The continuous time model was estimated using monthly data. Three discrete models were estimated, one each using monthly, quarterly and annual data. Our main results can be summarized as follows. First, the parameter estimates from the different models that we estimated are consistent with the Mundlak (1961)-Zellner (1968) hypothesis that temporal aggregation can account for slow speeds of adjustment in stock adjustment models. Second, we find that while the effects of temporal aggregation are substantial as we move from annual to quarterly to monthly specifications of the model, they are rather small when we move from the monthly to the continuous time specifications. This second result is consistent with findings in Christiano (1986b), where the length of the timing interval in a rational expectations model is treated as a free parameter. Christiano (1986b) plots the maximized value of the likelihood function of an annual data record against various values of the model timing interval. As the interval is reduced from an annual to a quarterly specification, the value of the likelihood function rises substantially. However, further decreases in the model timing interval result in smaller increases in the value of the likelihood function. This result is also consistent with findings in Christiano (1986a) in which a continuous time model of hyperinflation is estimated using monthly data. When an analogous discrete time model is fit to the same data, the results are virtually indistinguishable from the continuous time results.

The 11 free parameters of our continuous time model are

$$\Lambda^c = (r, a, b, c, e, A, \alpha, \beta, V_{11}, V_{22}, V_{12}).$$

Our discrete time model also has eleven free parameters:

$$\Lambda^d = (\phi, a, b, c, e, A, \rho, \mu, V_{11}^d, V_{22}^d, V_{12}^d).$$

Equation (40) implies that no more than nine parameters of the continuous time model can be identified. The same is true for the discrete time model. Consequently, we searched for a lower dimensional parameter set that was identified. We restricted our attention to sets that included $(\lambda-r)$ and Ψ for the continuous and discrete time models, respectively. For present purposes, it does not concern us that we cannot identify all the elements of Λ^c and Λ^d, since our principal motivation is to identify the adjustment speeds implied by the two models. These are controlled by $(\lambda-r)$ and Ψ in the continuous and discrete time cases, respectively. The parameter sets that we estimated are the following:

$$\zeta = (r, \alpha, \beta, \lambda-r, bc/a, (a+b+A)/a, V_{11}^c, V_{22}^c, V_{12}^c)$$

and

$$\xi = (\phi, \rho, \mu, \Psi, bc/a, (a+b+A)/a, V_{11}^d, V_{22}^d, V_{12}^d).$$

Christiano and Eichenbaum (1985) establish that ζ and ξ are identified.[8] In practice, we fixed the discount rates r and ϕ, a priori, at values which imply a monthly discount rate of .997.[9]

Both models were estimated using seasonally adjusted monthly data on nondurable manufacturing shipments and finished goods inventories. The data correspond to those used by Blinder (1986). These data are published by the Bureau of Economic Analysis (BEA), except that Blinder has converted BEA's end-of-month inventory stocks to beginning-of-month figures. We constructed quarterly and annual data by taking arithmetic averages of the monthly data. The data cover the period from February 1959 to April 1982 and are measured in millions of 1972 dollars. Shipments data are averages over the month. Means and trends were removed from the data using a second-order polynomial function of time and seasonal dummies.[10]

---

[8] Specifically, Christiano and Eichenbaum (1985) show that ς and ξ are <u>locally</u> identified. In addition, we show that, given any admissible ς, there are at least five other values of ς which are observationally equivalent, i.e., yield an identical value for the likelihood function. We constructed an algorithm to find these ς's in order to determine whether any of them is admissible in the sense of satisfying the nonnegativity conditions imposed by the model. Generally, we find that one other ς is admissible in this sense. This value of ς is obtained by exchanging the values of α and $(\lambda-r)$ and suitably adjusting r. As we point out later, our continuous time parameter estimates imply α = .082 and $(\lambda-r)$ = 5.29 with r = .003. This parameterization implies a relatively rapid speed of adjustment of actual to desired inventories. An alternative parameterization which yields the same value of the likelihood function is one in which α = 5.29 and $(\lambda-r)$ = .082. This implies that the speed of adjustment is very slow and there is relatively little serial correlation in the inventory holding cost shock. This parameterization can be ruled out as being implausible since it requires the discount rate to be r × 100 = 62,112 percent. We experimented with numerous parameterizations and always found that if we placed a reasonable upper bound on r, then global identification obtained. We found the same result regarding ς. In practice, we found that the likelihood function was not very sensitive to r. Consequently, the preceding identification arguments do not rule out the existence of alternative peaks which have values approximately similar to our estimated maximum.

[9] Our results were insensitive to the different values of r and β that we considered.

[10] This time trend can be rationalized as follows. Suppose that u(t) and v(t) are the sum of a covariance stationary component, as given by equation (18) and a linear function of time and seasonal dummies. Then the equilibrium laws of motion will have two components.

Table 1 reports the results of estimating the continuous time model using monthly data.[11] We are particularly interested in the implications of these estimates for the speed of adjustment statistics. The point estimate for $\lambda - r$ is 5.29 with 90 percent confidence interval given by (1.83, 8.75). This implies that

$T^c = 17 \ (10,49)$.

The 90 percent confidence interval is reported in parentheses. Thus the continuous time model implies that it takes 17 days to eliminate 95 percent of the gap between actual and desired inventories. This speed of adjustment seems plausible, especially in light of Feldstein and Auerbach's (1976) observation that even the largest swings in inventory stocks involve only a few days' worth of production.

We now turn to the results obtained with the discrete time models. Tables 2, 3, and 4 report results obtained with monthly, quarterly and annual data, respectively. The point estimates of Ψ obtained with monthly, quarterly and annual data are .14 (.036,.244), .28 (.070,.490) and .58 (.150,1.01), respectively. (Again, 90 percent confidence intervals are reported in parentheses.) The standard errors of the

---

The first component will be the law of motion given in the text. The second component will be a deterministic function of time and seasonal dummies. There are no restrictions across the two components. These claims are established in Christiano and Eichenbaum (1985). There are alternative ways to generate trend growth in inventories and sales. For example, the equilibrium laws of motion for $s(t)$ and $I(t)$ will inherit any unit roots in the VAR for $u(t)$ and $v(t)$. The fact that we choose to work with deterministic time trends does not necessarily reflect the view that this is the only reasonable model of trend growth for our variables. Instead, it reflects the fact that almost the entire empirical literature that we wish to address assumes the existence of deterministic time trends.

[11] In models where the timing interval is finer than the data sampling interval, estimates of the AR and MA parameters can be sensitive to the scale in which the data are measured. This contrasts with the case in which the timing interval coincides with the data sampling interval. In the latter case, multiplying the data by a constant scalar affects only the innovation variances, not the AR and MA parameters. To check that our continuous time speed of adjustment estimate is robust to a change of scale, we divided the data by 100 and reestimated the model parameters. The results were virtually unchanged.

estimates of Ψ increase with the degree to which the data are temporally aggregated. Presumably, this reflects the smaller number of data points that are available for the more temporally aggregated data.

The implied speed of adjustment statistics are given by

	Continuous	Monthly	Quarterly	Annual
Days to Close 95% of the Gap	17	46	212	1,980
Confidence Interval	(10,49)	(27,63)	(101,378)	(577,∞) [12]

The continuous time figures are repeated here for ease of comparison. The numbers in the last three columns of the first row correspond to T^d in (28)'. The number in the first column of the first row corresponds to T^c in (28). Numbers in parentheses in the second row are 90 percent confidence intervals.

Notice that the number of days required to close 95 percent of the gap between actual and desired inventories (T^d) is more than twice as large with monthly data, more than 12 times as large with quarterly data, and more than 115 times as large with annual data as the estimate obtained using the continuous time model. Evidently, the estimated speeds of adjustment are a monotonically decreasing function of the degree to which the data are temporally aggregated. We take this result to be supportive of the Mundlak-Zellner hypothesis that temporal aggregation can account for slow speeds of adjustment in stock adjustment models. The estimated adjustment speeds are plausible for the continuous time and monthly models, but implausibly slow--in our view--in the quarterly and annual models.

An interesting feature of our results is that the estimated speed of adjustment increases in diminishing increments as the model timing

---

[12] The upper bound of the ninety percent confidence interval for Ψ in the annual model is 1.01. This implies that firms never reach their target inventory level. This is why the reported upper bound of the ninety percent confidence interval for T^d in the annual model is ∞.

Table 1
Continuous Time Model
Monthly Data

Structural Parameters*		Reduced Form Parameters
α	.081 (.021)	$\theta_1^c = -1.85$
β	.082 (0.21)	$\theta_2^c = .851$
$\lambda - r$	5.29 (2.10)	$\theta_3^c = -.004$
bc/a	610.9 (9,120.5)	$c_1^c = \begin{bmatrix} -.772 & -.035 \\ -.032 & -.698 \end{bmatrix}$
$a/(a+b+A)$	0.00 (.001)	
$V = \begin{bmatrix} 13,244.5 & -507.3 \\ (12,046.2) & (4,854.8) \\ & 28,310.7 \\ & (25,150.5) \end{bmatrix}$		$c_2^c = \begin{bmatrix} -.104 & .009 \\ .088 & -.243 \end{bmatrix}$
		$c_3^c = \begin{bmatrix} -.002 & .003 \\ -.001 & .002 \end{bmatrix}$
		$V^c = \begin{bmatrix} 24,852.1 & 12,459.9 \\ & 187,924.0 \end{bmatrix}$
Log likelihood = -3,352.33		

* Standard errors are displayed in parentheses.

Table 2

Discrete Time Model

Monthly Data

Structural Parameters*		Reduced Form Parameters
μ	.910 (.027)	$\theta_1^d = -2.01$
ρ	.960 (.021)	$\theta_2^d = 1.14$
Ψ	.140 (.063)	$\theta_3^d = -.12$
bc/a	1.00 (1.17)	$C_1^d = \begin{bmatrix} -.910 & .008 \\ 0.00 & -1.10 \end{bmatrix}$
a/(a+b+A)	0.00 (.001)	
$V^d = \begin{bmatrix} 24{,}808.7 & 7{,}594.0 \\ (2{,}110.4) & (3{,}781.3) \\ & 54{,}792.8 \\ & (13{,}156.5) \end{bmatrix}$		$C_2^d = \begin{bmatrix} 0.00 & 0.00 \\ 0.00 & .130 \end{bmatrix}$

Log likelihood = -3326.97

* Standard errors are displayed in parentheses.

Table 3

Discrete Time Model

Quarterly Data

Structural Parameters*		Reduced Form Parameters	
μ	.824 (.077)	$\theta_1^d = -1.96$	
ρ	.854 (.063)	$\theta_2^d = 1.18$	
Ψ	.283 (.132)	$\theta_3^d = -.20$	
bc/a	.078 (.602)	$C_1^d = \begin{bmatrix} -.824 & -.007 \\ 0.00 & -1.14 \end{bmatrix}$	
a/(a+b+A)	0.00 (.001)		
$V^d = \begin{bmatrix} 65,530.8 & 8,337.6 \\ (9,731.9) & (14,396.0) \\ & 276,318.4 \\ & (41,021.8) \end{bmatrix}$		$C_2^d = \begin{bmatrix} 0.00 & 0.00 \\ 0.00 & .242 \end{bmatrix}$	
Log likelihood - = -1,161.52			

* Standard errors are displayed in parentheses.

Table 4

Discrete Time Model

Annual Data

Structural Parameters*		Reduced Form Parameters	
μ	.139 (.224)	θ_1^d =	-1.31
ρ	.584 (.256)	θ_2^d =	.500
Ψ	.584 (.525)	θ_3^d =	-.050
bc/a	.998 (.525)	$c_1^d = \begin{bmatrix} -.139 & -.038 \\ .206 & -1.17 \end{bmatrix}$	
a/(a+b+A)	.021 (.396)		
$V^d = \begin{bmatrix} 13,3765.1 & -42,203.5 \\ (42,042.6) & (60,428.8) \\ & 468,030.5 \\ & (146,721.4) \end{bmatrix}$		$c_2^d = \begin{bmatrix} 0.00 & 0.00 \\ -.029 & .333 \end{bmatrix}$	

* Standard errors are displayed in parentheses.

interval is reduced. The increase is very large, going from annual to quarterly data, but appears to have approximately converged at the monthly level. To see this, notice that the adjustment speed confidence intervals for the monthly and continuous time models overlap considerably. To investigate the conjecture that convergence has occurred with the monthly specification, we compared the discrete time reduced forms of the monthly and continuous time models.

The reduced forms of the continuous and discrete time models are reported in the second columns of Tables 1 and 2, respectively. These are similar along a number of interesting dimensions. First, C_3^c is close to zero, while the third-order term in $C^d(L)$ is exactly zero. Also, the (2,1)-elements of C_1^c and C_2^c are small and so compare well with the implication that sales fail to be Granger-caused by inventories.[13] One dimension along which the reduced forms differ concerns the first row of C_2^c, which does not appear to be close to zero. In contrast, the first row of C_2^d is identically equal to zero. Also, the variance of the second innovation error is three times larger in the continuous time model than in the discrete time model. Unfortunately, the importance of these differences and similarities is hard to judge, since we do not have the relevant distribution theory. Moreover, it is not clear that a direct comparison of the reduced form parameters is the most revealing one.

In our view, it is more interesting to compare the implications of the two reduced forms for both sets of structural parameters. We are particularly interested in the implications of the reduced form representation of the data emerging from the continuous (discrete) time model for the structural parameters of the discrete (continuous) time model. Consider first the implications of the reported reduced forms for the structural parameters of the continuous time model. Since the continuous time model is identified, the reduced form parameters in the second column of Table 1 map uniquely into the parameter values reported in the first column of Table 1. It is less obvious how to deduce the implications of the reduced form emerging from the discrete time model for the structural parameters of the continuous time model.

---

[13] We noted in Section 2 that this assumption is frequently made in the inventory literature.

Since the reduced form of the discrete time model does not satisfy the cross equation restrictions implied by the continuous time model, there is in fact no set of continuous time structural parameters consistent with the discrete time model reduced form. In view of this, we decided that the most sensible thing to do was to compute the set of continuous time parameters that comes closest to reproducing the discrete time reduced form in Table 2.

A natural candidate for this set of parameters is the probability limit of the maximum likelihood estimator of the continuous time structural parameters calculated under the assumption that the data are generated by the estimated reduced form corresponding to the discrete time model.[14] If the discrete time model is true, then the estimates of the continuous time model obtained using monthly data ought to be close to this probability limit. These probability limits are reported in the second of the two columns labeled "Plim" in Table 5. Numbers in parentheses are the estimated parameter values taken from the first column of Table 2. We find some discrepancies. For example, the plim of α is .035, while its estimated value is .081. Other discrepancies which stand out are the results for bc/a, V_{22} and V_{12}. Unfortunately, we cannot draw any definitive conclusions regarding the magnitude of these differences in the absence of the relevant distribution theory. Nevertheless, it is interesting to note the similarity between the estimated value of $\lambda - r$ and its reported probability limit. As noted earlier, the estimated value of $\lambda - r$ implies that firms close 95 percent of the gap between actual and desired inventories in 17 days. The estimated probability limit of this number under the assumption that the data are generated by the discrete time monthly model is 19.5 days.

We now consider the implications of the two reduced form representations for the structural parameters of the discrete time model. In the first "Plim" column of Table 5 we report the

---

[14] These were computed by maximizing the frequency domain approximization to the Gaussian likelihood function in which the periodogram was replaced by the spectral density function implied by the reduced form parameters in Table 2. The justification for calling the resulting numbers "probability limits" is given in Christiano (1984), where this technique is applied in another context.

Table 5
Probability Limits

Discrete Time Model Parameter	Plim[1]	Continuous Time Model Parameter	Plim[2]
ρ	.940 (.960)	α	.035 (.081)
μ	.938 (.910)	β	.164 (.082)
Ψ	.116 (.140)	$\lambda - r$	4.60 (5.30)
bc/a	51.45 (1.00)	bc/a	.879 (611.1)
a/(a+b+A)	0.00 (0.00)	a/(a+b+A)	0.00 (0.00)
V_{11}^d	24,951.7 (24,808.7)	V_{11}	19,013.6 (13,244.5)
V_{22}^d	200,570.0 (54,792.8)	V_{22}	8,276.7 (28,310.7)
V_{12}^d	11,701.0 (7,594.0)	V_{12}	2,661.6 (-507.3)

[1] Probability limit of parameters of monthly discrete time model, assuming data are generated by reduced form in Table 1. Numbers in parentheses are parameter estimates obtained from the data and reported in Table 1.

[2] Probability limit of parameters of continuous time model, assuming data are generated by reduced form in Table 2. Numbers in parentheses are parameter estimates obtained from the data and reported in Table 2.

probability limits of the structural parameters of the discrete time monthly model. These were calculated under the assumption that the data are generated by the continuous time model. If the continuous time model is true, then the estimates of the structural parameters of the discrete time model obtained using the monthly data ought to be close to the corresponding probability limits reported in Table 5. In fact, these appear to be quite close to each other. The principal discrepancy is that bc/a is larger than the value reported in Table 2. In addition, V_{22}^d and V_{12}^d are somewhat different from the values reported in Table 2. As before, we cannot draw any definitive conclusions from this exercise without the relevant distribution theory. Nevertheless, it is interesting to note how similar the estimate of Ψ reported in Table 2 is to its plim in Table 5. In particular, inferences about the speed of adjustment of actual to target inventories are basically the same for the two values of Ψ.

We conclude from the results in Table 5 that, when viewed from the point of view of their implications for the discrete time parameters, the reduced forms in Tables 1 and 2 are fairly similar. Some differences are apparent when examined from the point of view of certain structural parameters of the continuous time model.

A third way to compare the two reduced form representations is to compare their log likelihood values. The difference between the log likelihood value of the discrete time monthly and continuous time models is equal to 25.36. In this sense, the discrete time monthly model fits the data better than the continuous time model. On the other hand, the likelihood ratio statistic obtained when either of the two models is compared with an unrestricted reduced form ARMA(3,3) model indicates rejection of both structural models at essentially the same level. The log likelihood value of the unrestricted ARMA(3,3) model is 3,307.5, which is significantly greater than the log likelihood values associated with both the continuous and discrete time monthly models (see Tables 1 and 2).

Overall, we conclude that the monthly discrete time and continuous time models appear to be fairly similar when examined from the perspective of the reduced form time series representations that they imply for the monthly data. Next, we report some diagnostic tests on the underlying statistical adequacy of the two structural models.

The validity of the formulas used to compute the confidence intervals around our speed of adjustment estimates requires that the underlying models be correctly specified. Unfortunately, we found evidence against this hypothesis. As we indicated, a likelihood ratio test rejects both models against an unrestricted ARMA (3,3) alternative. We also computed the multivariate Box-Pierce statistics proposed by Li and McLeod (1981) to test for serial correlation in the fitted residuals from the continuous time and monthly discrete time models. These statistics were computed at lags 12 and 24 and are denoted by BP(12) and BP(24), respectively. Under the null hypothesis that the underlying disturbances are white noise, BP(k) is drawn from a chi-square distribution with $4 \times k - n$ degrees of freedom, where n is the number of free parameters.[15] In our case, $n = 9$. The Box-Pierce statistics for the continuous time model are BP(12) = 162 and BP(24) = 278. For the discrete time model, they are BP(12) = 386 and BP(24) = 602. These statistics indicate a substantial departure from white noise in the fitted residuals. Because the likelihood ratio statistics and Box-Pierce statistics supply evidence against our models, the speed of adjustment confidence intervals that we reported above must be interpreted with caution.

To what extent are our results sensitive to the way in which we specified our discrete time model? As we indicated in Section 3, there are at least two ways to choose a discrete time analogue to the continuous time model of Section 2. Our procedure was to specify the shocks in the discrete time model to have the same representation as the point-in-time sampled representations of the continuous time shocks. Since our continuous time shocks are AR(1), this implies an

---

[15] Li and McLeod (1981) derive the distribution for their test statistic under the assumption that the model being estimated is an unconstrained vector ARMA with independent, identically distributed disturbances. They show that BP(k) has an asymptotic chi-square distribution with $m^2 k - \ell$ degrees of freedom, where m is the number of equations in the vector ARMA model and ℓ is the number of AR and MA parameters. We assume that the appropriate modification regarding the number of degrees of freedom in our problem is obtained by replacing ℓ by n.

AR(1) representation for the shocks in the discrete time model. We adopted this specification of the discrete time model because it matches well with what is commonly done in the literature.[16] An alternative would have been to specify the shocks in the discrete time model so as to produce a reduced form for that model with AR and MA orders identical to those implied by the continuous time model. This can be accomplished by adding a first-order moving average term to the shocks in the discrete time model. We conjecture that the effect of these moving average terms would be to raise the estimated speed of adjustment implied by the discrete time model. This conjecture is based on the belief that the additional MA terms would take over some of the burden borne by the AR parameters--one of which controls the speed of adjustment--for accommodating the serial correlation in the data. This would be consistent with results in Telser (1967). As yet, we have not formally investigated this conjecture. However, it is important to note that these comments illustrate the observations made in Christiano and Eichenbaum (1987, Section 2B), where we argued that the temporal aggregation effects of shrinking the model's timing interval can have the same effect on the reduced form implications of a model as allowing for more serial correlation in the unobserved shock terms.

---

[16]See for example, Blinder (1986), Eichenbaum (1984), Maccini and Rossana (1984) and the references in McCallum (1984).

References

Altonji, J. G. (1986): "Intertemporal Substitution in Labor Supply: Evidence From Micro Data," manuscript.

Blanchard, Olivier J. (1983): "The Production and Inventory Behavior of the American Automobile Industry," Journal of Political Economy, 91, 365-400.

Blinder, Alan S. (1981): "Inventories and the Structure of Macro Models," American Economic Review, 71, 11-16.

Blinder, Alan S. (1986): "Can the Production Smoothing Behavior Model of Inventory Behavior Be Saved," manuscript.

Blinder, Alan S. and D. Holtz-Eakin (1984): "Inventory Fluctuations in the United States Since 1929," manuscript.

Bryan, William R. (1967): "Bank Adjustments to Monetary Policy: Alternative Estimates of the Lag," American Economic Review, 57, 855-864.

Christiano, Lawrence J. (1984): "The Effects of Aggregation Over Time on Tests of the Representative Agent Model of Consumption," manuscript, Federal Reserve Bank of Minneapolis.

Christiano, Lawrence J. (1986a): "A Continuous Time Model of Cagan's Model of Hyperinflation Under Rational Expectations," manuscript, Federal Reserve Bank of Minneapolis.

Christiano, Lawrence J. (1986b): "A Method for Estimating the Timing Interval in a Linear Econometric Model, with an Application to Taylor's Model of Staggered Contracts," Journal of Economic Dynamics and Control, 9, 363-404.

Christiano, Lawrence J. and Martin Eichenbaum (1985): "A Continuous Time, General Equilibrium, Inventory-Sales Model," manuscript, Federal Reserve Bank of Minneapolis.

Christiano, Lawrence J. and Martin Eichenbaum (1987): "Temporal Aggregation and Structural Inferrence in Macroeconomics," in Bubbles and Other Essays, ed. by Karl Brunner and Alan H. Meltzer, Carnegie-Rochester Conference Series on Public Policy, vol. 26, Spring, Amsterdam: North-Holland.

Dimelis, Sophia P. and Tryphon Kollintzas (1988): "A Linear Rational Expectations Equilibrium Model of the American Petroleum Industry," Chapter IV in this book.

Eckstein, Zvi and Martin S. Eichenbaum (1985): "Quantity-Constrained Equilibria in Regulated Markets: The U.S. Petroleum Industry, 1947-1972," in Energy, Foresight and Strategy, ed. by Thomas J. Sargent, Washington, D.C.: Johns Hopkins University Press.

Eichenbaum, Martin S. (1984): "Rational Expectations and the Smoothing Properties of Inventories of Finished Goods," *Journal of Monetary Economics*, 14: 71-96.

Eichenbaum, Martin S., Lars P. Hansen, and S. Richard: (1985): "The Dynamic Equilibrium Pricing of Durable Goods," manuscript.

Feldstein, Martin S. and Alan Auerbach (1976): "Inventory Behavior in Durable Manufacturing: The Target Adjustment Model," *Brookings Papers on Economic Activity*, 2, 351-408.

Goodfriend, Marvin (1985): "Reinterpreting Money Demand Regressions," in *Understanding Monetary Regimes* ed. by Karl Brunner and Allan H. Meltzer, Carnegie-Rochester Conference Series on Public Policy, vol. 22, Amsterdam: North-Holland.

Hannan, Edward J. (1970): *Multiple Time Series*, New York: Wiley.

Hansen, Gary D. (1985): "Indivisible Labor and the Business Cycle," *Journal of Monetary Economics*, 16: 309-327.

Hansen, Lars P. and Thomas J. Sargent (1980a): "Methods for Estimating Continuous Time Rational Expectations Models from Discrete Data," Research Department, Staff Report 59, Federal Reserve Bank of Minneapolis.

Hansen, Lars P. and Thomas J. Sargent (1980b): "Formulating and Estimating Dynamic Rational Expectations Models," *Journal of Economic Dynamics and Control*, 2: 351-408.

Hansen, Lars P. and Thomas J. Sargent (1981): "Formulating and Estimating Continuous Time Rational Expectations Models from Discrete Data," manuscript.

Li, W. K. and A. I. McLeod (1981): "Distribution of the Residual Autocorrelations in Multivariate ARMA Time Series Models," *Journal of the Royal Statistical Society* (Series B), 43, 231-239.

Lucas, Robert E., Jr. and Edward C. Prescott (1971): "Investment Under Uncertainty," *Econometrica*, 39, 659-681.

Luenberger, David G. (1969): *Optimization by Vector Space Methods*, New York: Wiley.

Maccini, Louis J. and Robert J. Rossana (1984): "Joint Production, Quasi-Fixed Factors of Production, and Investment in Finished Goods Inventories," *Journal of Money Credit and Banking*, 16, 218-236.

MaCurdy, Thomas E. (1981): "An Empirical Model of Labor Supply in a Life-Cycle Setting," *Journal of Political Economy*, 89, 1059-1085.

McCallum, Bennett T. (1984): "Inventory Fluctuations and Macroeconomic Analysis: A Comment," manuscript.

Mundlak, Yair (1961): "Aggregation Over Time in Distributed Lag Models," International Economic Review, 2, 154-163.

Rogerson, R. (1984): "Indivisible Labor, Lotteries and Equilibrium," manuscript.

Sargent, Thomas J. (1979): Macroeconomic Theory, New York: Academic Press.

Telser, L. G. (1967): "Discrete Samples and Moving Sums in Stationary Stochastic Processes," Journal of the American Statistical Association, 62, 484-499.

West, Kenneth D. (1986): "A Variance Bounds Test of the Linear-Quadratic Inventory Model," Journal of Political Economy, 94, 374-401.

Zellner, Arnold (1968): "Note on Effect of Temporal Aggregation on Estimation of Stock Adjustment Equation," manuscript.

CHAPTER IV

A LINEAR RATIONAL EXPECTATIONS EQUILIBRIUM MODEL OF THE AMERICAN PETROLEUM INDUSTRY[1]

Sophia P. Dimelis
Athens School of Economics and Business

Tryphon Kollintzas
University of Pittsburgh

Abstract

This paper develops and estimates a model of the American petroleum industry. The model accounts for the storable and exhaustible nature of petroleum as well as the strategic interaction of agents operating in the stochastic environment of the markets for crude and refined petroleum products. The linear rational expectations modelling framework is adopted. The formulation and the econometric specification of the model are motivated by a statistical and vector autoregression analysis of the annual data for the post World War II period. The parameter estimates of the model over that period conform to the model's restrictions. In addition, the overall fit of the model judged from the usual diagnostic statistics seems to be relatively good. Important findings of this empirical test are: a not too inelastic domestic demand for refined petroleum products; a marginal cost of domestic crude petroleum production that is an increasing function of cumulative production (exhaustibility), evidence of production smoothing inventory behavior; fast inventory adjustment to desired inventory levels; and finally, upward sloping foreign supplies of crude and refined petroleum products.

1. Introduction

In the last fifteen years an extensive literature aiming at explaining the behavior of the American petroleum industry has been

---

[1] We are grateful to Alan Blinder, James Cassing, Shirley Cassing, Martin Eichenbaum, Dennis Epple, Edward Green, Herbert Mohring, Edward Prescott, and Kyprianos Prodromides for comments. Part of this work was completed when the second author was visiting the Athens School of Economics and Business and the Federal Reserve Bank of Minneapolis. He would like to thank these institutions and their staff for their support and hospitality.

developed.² That is, the time series behavior of production, imports, sales, inventories, and prices of crude and refined petroleum products. However, this literature has not been very successful in, perhaps, its most important task - the provision of a model that is appropriate for predicting this behavior and evaluating the effects of any one of several government policies that affect it. This is because the foreign and domestic demands and supplies of the petroleum industry in existing models are either specified in an ad hoc manner or, although derived from the purposeful behavior of the economic agents involved, fail to capture the dynamic and strategic features characterizing this behavior. Ad hoc specifications do not allow for the interpretation of the estimated parameters. Consequently, as Lucas (1976) observed, although these models may fit the historical data well, they are not appropriate for prediction and policy evaluation purposes.³ On the other hand, given that petroleum products are typically both exhaustible and durable, and since the extraction, processing and distribution of these products involve investments that are both irreversible and subject to adjustment costs, all economic participants in these markets have strong incentives to care about the future paths of prices and costs. Moreover, since petroleum markets are characterized by a number of agents whose decisions are mutually interrelated, it is appropriate to account for this interdependence and interaction of agents. For those reasons the petroleum industry cannot be studied adequately in frameworks that ignore the dynamic and strategic features underlying the behavior of the economic agents involved.⁴

---

²Important contributions in this literature are: Houthakker and Taylor (1970), Adelman (1972), Adams and Griffin (1972), Nordhaus (1974), Mitchell (1976), Kennedy (1976), Verleger (1982a, 1982b), Chao and Manne (1982), and Hubbard and Weiner (1983). Petroleum industry models have also been included in large macroeconomic models. See, e.g., Pierce and Enzler (1974), Perry (1975), Eckstein (1978, 1981), Fair (1978), Klein (1978), Hudson and Jorgenson (1978), and Mork and Hall (1980a, 1980b, 1981).

³See also Nerlove (1972) and Sargent (1981).

⁴This has been pointed out by a number of authors. See, e.g., Adelman (1972) and Mitchell (1976).

A general framework for studying behavior in such industries was put forward by Sargent (1985) and is known as "rational expectations modelling." Borrowing Sargent's expression, "this analysis requires two things: first, tools that permit analyzing individual agent's choices of intertemporal strategies which are constrained both by the physical technologies and by the intertemporal strategies chosen by the other market participants, both private and governmental; and second, an equilibrium concept suitable for studying the dynamic interaction of a collection of agents, which ensures that agent's choices of intertemporal strategies are mutually consistent with the physical constraints and with their perceptions of each other's strategies." Although the contributors in Sargent's (1985) volume applied this methodology to the study of several segments of the petroleum industry, to our knowledge there has been no attempt to apply this methodology in studying an integrated model of the petroleum industry.[5]

The purpose of this paper is to develop and estimate an integrated model of the American petroleum industry following the principles laid out above. In particular, the model employed is a Linear Rational Expectations Model (LRE). The advantage of using LRE modelling is that it gives estimable functional forms. The disadvantage of using LRE is that a very tight structure is imposed on the data.[6]

In Section 2 we present some stylized facts of the American petroleum industry. In Section 3 we develop the model. Section 4 yields the model's solution - the linear rational expectations equilibrium laws of motion in the crude and refined petroleum product markets. Section 5 presents estimates and tests of the model using

---

[5] Aiyagari and Riezman (1985) model the market for a storable resource as a dynamic game between a dominant seller and the buyers of the resource, abstracting from exhaustibility. Eckstein and Eichenbaum (1985a) model an economy that faces a fixed supply of imported oil. Eckstein and Eichenbaum (1985b) model the refined petroleum products market given the demand for refined petroleum products. Hansen, Epple and Roberds (1985) model the market for an exhaustible resource as several dynamic games between suppliers given the demand for the resource. Epple (1985) tests the exhaustible resource supply model developed in Epple and Hansen (1981) on data of the American crude petroleum and natural gas markets.

[6] See Blanchard (1983) for an excellent presentation of the pros and cons of LRE modelling in a similar setup.

annual data for the post World War II period. Section 6 concludes. Technical material is delegated to an appendix. Another appendix contains the data and their sources.

2. **Stylized Facts**

2.1 <u>Facts and Figures</u>

This section describes the actual behavior of consumption, production, imports, inventories, and prices in the American petroleum industry since 1947. It then presents and analyzes some basic statistical facts associated with these variables in order to obtain the appropriate intuition in formulating and estimating the model.

There are, of course, many different kinds of crude and refined petroleum products. But in this study we have chosen to consider crude and refined petroleum products as composite homogeneous goods, denoted by C and R, respectively.[7] Tables 1 and 2 present and Figures 1 and 2 illustrate the actual behavior of consumption, production, imports, inventories, and prices of C and R. The most striking feature of the data is that behavior prior to the early seventies is in sharp contrast to behavior since then. Smooth co-movements prior to the early seventies seem to break down since then. Since there were a number of major institutional and other environmental changes during the early seventies, it is apparent that a model of this industry must allow for channels through which such changes in the environment could account for, at least in part, this feature of the data.

The trend in domestic consumption of crude and refined petroleum products has been upward. However, there have been a number of significant downward movements during recessions. As Figures 1 and 2 illustrate these downward movements were especially profound during the recessions of 1975 and 1981-82. Domestic production of crude petroleum products moved very much like the domestic consumption of these products until the late sixties. But since then, their movements have been dissimilar with the gap between them getting bigger in the seventies and closing somewhat in the early eighties. Imports of crude

---

[7] See Appendix B for the composition of C and R.

petroleum have more than bridged this gap since crude petroleum inventories have also been on an upward trend. There were some major government policy changes that might have contributed to this behavior. Most importantly, the quota on crude petroleum that was in effect from March 1959 to April 1973 and the price controls that were in effect from August 1971 to October 1981.[8] Thus, as the quota was keeping imports away, the domestic real price of crude petroleum exceeded the world real price (i.e., price at which crude petroleum was imported) until 1973, and, as the quota was abolished and the price controls were initialized, the domestic price fell below the world price. Since a model with a quota turns out to be a very different model from one where there is no quota, these observations justify the considerable amount of complication introduced in the model by actually deriving two equilibrium laws of motion in the markets for C and R - with and without a quota - and by having to switch between these laws of motion during estimation. Another possible determinant of the relative drop in domestic production could be that domestic production costs increased faster than foreign costs. Later, we will try to account for this by making domestic production costs dependent on the cumulative production of crude petroleum (the Epple and Hansen (1981) exhaustible resource model).

The real world price of oil was falling until 1973, despite the declared intentions of OPEC (formed in 1960) to raise them. In the aftermath of the October 1973 Arab-Israeli War, OPEC reduced production and its Arab members declared an embargo on exports to the United States and The Netherlands that lasted until March 1974 and July 1974, respectively. Between 1970 and 1974 prices of crude petroleum products quadrupled. Also, during the Iranian crisis of 1978-1979 the production of that major petroleum exporting country was severely curtailed. From 1978 to 1979 prices of crude petroleum products doubled, surpassing even OPEC's posted prices.

---

[8] For details of the oil import programs see Bohi and Russel (1976), while an extensive analysis of the price controls is given in Kalt (1982).

TABLE 2.1: Crude Petroleum Market: Data Estimates, 1947 to 1984

YEAR	Inventories CINV	Domestic Production COUT	Net Imports CIMP	Refiners Consumption CONR	Domestic Real Price CDPRID	World Real Price CFPRID
	Millions of barrels				Dollars per barrel	
1947	235.70	1,989.90	51.10	2,041.70	3.90	3.59
1948	235.00	2,167.30	89.40	2,229.50	4.92	4.55
1949	262.20	1,999.20	120.60	2,121.80	4.90	4.52
1950	260.20	2,155.70	142.90	2,302.90	4.80	4.25
1951	255.90	2,452.70	150.50	2,595.10	4.53	4.00
1952	264.00	2,513.70	182.90	2,680.90	4.46	3.95
1953	279.70	2,596.20	216.60	2,807.70	4.71	3.96
1954	284.80	2,567.60	225.90	2,805.90	4.79	4.08
1955	272.40	2,766.30	273.80	3,033.30	4.69	4.01
1956	279.20	2,910.50	313.20	3,216.30	4.57	4.09
1957	286.60	2,912.10	323.10	3,218.40	4.87	4.24
1958	303.40	2,744.20	343.70	3,105.80	4.80	4.04
1959	285.50	2,895.70	349.80	3,249.00	4.57	3.72
1960	282.00	2,915.80	368.50	3,297.60	4.44	3.56
1961	268.70	2,983.70	378.30	3,348.90	4.41	3.58
1962	281.80	3,049.00	409.20	3,456.60	4.34	3.46
1963	283.40	3,153.70	411.00	3,577.00	4.26	3.42
1964	271.10	3,209.30	437.20	3,651.80	4.18	3.33
1965	265.80	3,290.10	450.90	3,750.60	4.08	3.25
1966	256.20	3,496.50	445.60	3,919.50	3.98	3.17
1967	278.80	3,730.20	385.10	4,049.10	3.91	3.10
1968	345.00	3,882.70	499.90	4,356.50	3.77	2.94
1969	371.10	3,956.30	551.50	4,510.20	3.75	2.84
1970	368.70	4,129.60	517.60	4,633.50	3.59	2.78
1971	382.40	4,077.80	658.10	4,751.90	3.64	2.84
1972	366.40	4,103.70	856.60	4,979.50	3.52	2.86
1973	347.20	4,006.00	1,233.50	5,237.20	3.68	3.35
1974	349.50	3,831.80	1,312.30	5,115.00	5.69	10.55
1975	378.60	3,666.50	1,509.10	5,169.10	6.04	9.91
1976	385.10	3,577.20	1,943.80	5,502.00	5.92	10.12
1977	404.10	3,618.10	2,407.30	5,960.10	6.05	10.12
1978	469.40	3,769.60	2,272.00	6,018.00	6.17	9.52
1979	493.00	3,715.50	2,315.20	5,961.40	7.12	12.16
1980	562.30	3,738.20	1,841.30	5,466.90	9.64	18.32
1981	674.90	3,734.10	1,571.00	5,208.90	12.73	18.69
1982	771.10	3,741.80	1,266.10	4,977.30	10.96	16.14
1983	801.70	3,759.20	1,257.90	4,934.40	9.78	13.71
1984	884.40	3,819.10	1,292.20	5,061.70	9.28	12.87

Source: See Appendix B.

Note: Real prices are nominal prices deflated by the implicit GNP deflator.

Crude Petroleum Time Series
Annual Data: 1947-1984

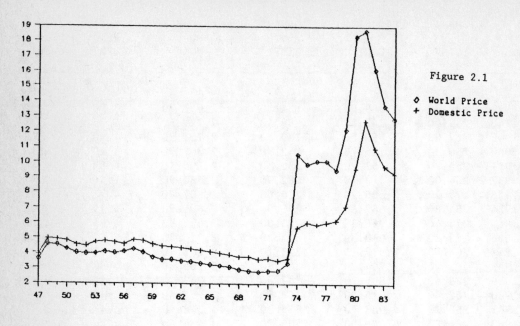

Figure 2.1

◇ World Price
+ Domestic Price

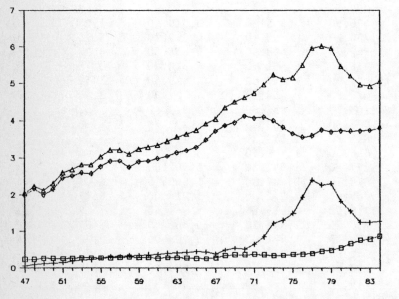

Figure 2.2

◇ Domestic Production
△ Consumption
□ Inventories
+ Imports

TABLE 2.2: Market of Refined Petroleum Products: Data Estimates, 1947-1984

YEAR	Inventories RINV	Domestic Production ROUT	Net Imports RIMP	Refiners Consumption RSAL	Domestic Real Price RDPRID	World Real Price RFPRID
	Millions of barrels				Dollars per barrel	
1947	224.40	1,865.40	-56.20	1,878.20	5.93	3.06
1948	211.60	2,060.80	-35.80	2,002.10	7.12	4.29
1949	270.30	1,994.20	-4.40	1,990.30	6.44	3.18
1950	274.20	2,164.30	56.00	2,179.30	6.55	3.23
1951	259.20	2,452.80	3.70	2,412.10	6.61	3.14
1952	299.90	2,556.20	7.40	2,535.40	6.48	3.21
1953	320.70	2,650.10	14.30	2,611.80	6.63	3.12
1954	359.00	2,650.00	28.40	2,654.20	6.49	3.36
1955	354.80	2,861.30	47.50	2,855.00	6.48	3.64
1956	361.10	3,027.10	55.00	2,964.20	6.59	3.96
1957	424.00	3,006.50	44.40	2,963.20	6.82	4.51
1958	467.30	2,939.50	176.30	2,979.00	6.12	4.08
1959	427.80	3,088.10	222.70	3,059.00	5.99	3.53
1960	456.90	3,133.50	221.70	3,142.40	5.85	3.43
1961	448.00	3,184.70	257.80	3,148.60	5.90	3.48
1962	484.10	3,286.20	289.20	3,287.90	5.74	3.37
1963	482.40	3,344.90	287.90	3,330.60	5.62	3.24
1964	496.70	3,422.70	315.60	3,422.80	5.25	3.14
1965	496.60	3,517.70	381.50	3,514.50	5.32	3.10
1966	499.80	3,676.30	421.10	3,654.30	5.34	2.92
1967	521.80	3,931.00	428.80	3,873.00	5.28	2.83
1968	579.80	4,143.50	455.00	4,115.10	4.95	2.76
1969	608.20	4,269.90	519.30	4,285.20	4.74	2.55
1970	592.90	4,396.00	636.00	4,372.40	4.78	2.47
1971	616.50	4,552.00	693.00	4,511.40	4.79	2.64
1972	657.10	4,734.90	797.30	4,800.20	4.64	2.69
1973	591.80	5,001.00	965.60	4,954.10	5.29	3.51
1974	638.70	4,870.20	838.20	4,832.50	9.04	10.10
1975	676.40	4,944.60	624.90	4,891.90	9.41	9.48
1976	729.10	5,246.10	651.00	5,290.00	9.49	9.18
1977	685.20	5,562.40	718.80	5,426.90	9.96	10.04
1978	820.70	5,608.90	648.60	5,669.00	9.53	9.06
1979	760.60	5,524.80	599.50	5,524.80	12.30	12.34
1980	760.60	5,130.40	488.20	5,166.40	17.10	16.24
1981	724.60	4,939.60	400.30	4,967.80	18.88	16.94
1982	696.40	4,791.30	302.80	4,876.50	16.73	15.03
1983	611.20	4,732.60	316.00	4,789.20	14.46	14.26
1984	554.60	4,963.50	413.30	4,911.00	13.45	13.92

Source: See Appendix B.

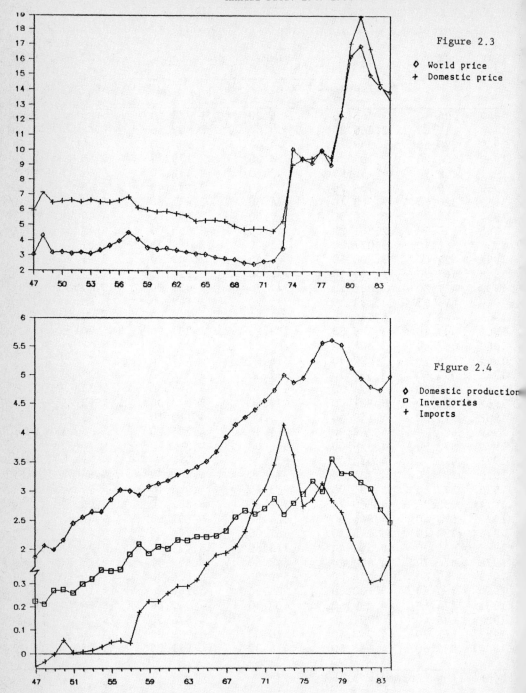

Refined Petroleum Products Time Series
Annual Data: 1947-1984

Figure 2.3

◊ World price
+ Domestic price

Figure 2.4

◊ Domestic production
□ Inventories
+ Imports

TABLE 2.3:   Basic Statistics: 1947 to 1984

Variable*	Mean	Standard Deviation	Coefficient of variation %
CINV	369.66	162.83	44.04
COUT	3,253.33	640.13	19.67
CONR	4,007.97	1,186.45	29.60
CIMP	773.01	691.37	89.43
CDPRID	5.40	2.20	40.74
CFPRID	6.30	4.65	73.80
RINV	511.71	165.98	32.43
ROUT	3,795.39	1,131.42	29.81
RSAL	3,785.32	1,137.45	30.04
RIMP	348.17	279.11	80.16
RDPRID	7.84	3.73	47.57
RFPRID	13.94	4.49	75.84

* See Appendix B for more details on the symbols and sources of variables.

As Table 2.3 makes clear, then, the real world price of crude petroleum (CFPRID) and imports (CIMP) have been very variable on an absolute scale and considerably more variable than any of the other variables of interest. These observations suggest that the real world price and imports of crude petroleum have been subject to shocks that cannot be modelled by simple ARMA laws of motion and justify our use of intervention analysis in modelling these shocks (see Kollintzas and Geerts (1984)).

Refined petroleum product variables have followed similar patterns as their crude petroleum counterparts with a notable exception. Inventories of refined petroleum products have been declining since the late seventies. President Carter's Strategic Petroleum Reserve Program which stored about four hundred million barrels of crude petroleum between 1978 and 1984 (or about forty-five percent of the private sector's inventories at the end of 1984), might have contributed to this. This suggests that the Strategic Petroleum Reserve should be treated as a distinct component of inventories that follows a different law of motion than the private sector's inventories.

Also, it is important to note that the coefficients of variation of the production of crude and refined petroleum products are smaller than the coefficients of variation of sales of crude and refined petroleum products, respectively.[9] This suggests that one of the motives for holding inventories is production smoothing. Moreover, it suggests that this motive is more important than the backlog cost avoidance motive.[10] Further, the speculative precaution motive of holding inventories may have been behind the increase of inventories immediately after the 1973-74 and 1978-79 periods.[11] This observation will be used later to justify our assumption that other than for speculation/precaution purposes, refineries hold inventories for production smoothing purposes (a synthesis of the Blinder (1982) output inventory model with the Kollintzas and Husted (1984) input inventory model).[12]

2.2 Vector Autoregressions

To further investigate the petroleum markets, a number of multivariate Granger-causality tests were performed. Although these

---

[9] Blinder (1984) presents evidence from seasonally adjusted monthly data that the variance of production is equal to the variance of sales. West (1986) presents evidence from seasonally adjusted monthly data that the variance of production is somewhat greater than the variance of sales. Since their data are in dollars and seasonally adjusted, their findings should not be taken to contradict ours. In fact, production smoothing should be underestimated in seasonally adjusted data as long as sales tend to be more seasonal than production. Moreover, the dollar and more aggregate (C and R are combined) figures of Blinder and West are more likely to be subject to measurement errors than are our barrel figures for C and R.

[10] In the absence of productivity shocks, the variance of production can be greater than the variance of sales if the backlog cost avoidance motive is sufficiently strong to dominate the production smoothing motive. See Blanchard (1983). However, in the presence of productivity shocks even without a backlog cost avoidance motive it is possible that the variance of production exceeds that of sales. See Blinder (1984).

[11] Evidence for the speculative motive for holding inventories is given in the works of Verleger (1980b, 1982a) and Hubbard and Wiener (1983).

[12] Strictly convex adjustment costs provide another incentive for production smoothing. See Eichenbaum (1983, 1984) and Blanchard (1983).

conclusions are only "informative" of the direction of causality and very sensitive to data transformations and test methodology, they can be useful in providing some intuition about model selections.

Thus, four different systems of vector autoregressions (VAR) were estimated using annual data from the entire period under consideration (1947-1984). Each system includes the same number of two-lag variables and the marginal significance of excluding the lags of each of the variables in the system is tested using the F-statistic. All variables in the VARs were either first differenced or detrended to remove nonstationarity.[13] In every equation of the system, the null hypothesis of zero coefficients of the two lags of each right hand variable is tested and the values of F along with their marginal significance (MS) levels are reported in each Table. MS levels between 0.05 and 0.075 (0 and 0.05) were taken to suggest week (strong) Granger-causality.

Table 2.4 reports the results of the various exclusions tests conducted on a VAR system including the series: crude petroleum domestic real price (CDPRD), inventories (CINV), domestic output (COUT) and imports (CIMP) respectively, and finally, the real GNP (RGNP).[14]

It is interesting to notice from Table 2.4 that CDPRD is not Granger-caused by any relevant variable in the system (i.e., MS >.075). This result can be explained by reference to the various price controls that were effective for most of the period under consideration as analyzed in the previous paragraphs of this section. Futhermore, imports (CIMP) do not seem to Granger-cause any other relevant variable, but are Granger-caused by RGNP and lagged imports. This again appears to be consistent with the regulatory import policies which limited imports for many years and thus insulated the domestic

---

[13] There is confusion in the literature as to whether first differencing should be preferred to detrending as a method of removing nonstationarity as required for the application of the asymptotic F-Statistic in Granger causality tests. For more details on the effects of each method on the causality relations see Kang (1985), Nelson and Kang (1981, 1984), Nelson and Plosser (1982), and Mankiw and Shapiro (1985) among others.

[14] For more details on the definitions, sources, and units of measurement of these and subsequent variables see Appendix B.

economy from the world markets. Similar conclusions were drawn by

TABLE 2.4

Granger-Causality Tests with specified variables in the Crude Petroleum Industry

Equation	CDPRD F	CDPRD MS	CINV F	CINV MS	COUT F	COUT MS	CIMP F	CIMP MS	RGNP F	RGNP MS
CDPRD	.007	.99	1.73	.19	1.99	.16	1.40	.26	1.27	.29
CINV	4.43	.02	8.56	.002	3.49	.05	2.50	.10	6.01	.008
COUT	.24	.79	.09	.91	.36	.70	.24	.79	.03	.97
CIMP	.79	.46	.45	.64	.14	.87	5.58	.01	4.14	.03
RGNP	10.26	.0006	2.19	.13	1.45	.25	1.36	.27	1.62	.22

Note: All variables were first-differenced and a constant was included in each equation.

Eckstein and Eichenbaum (1985b) for the period 1947-1973 using quarterly data. This again justifies our subsequent choice of a two track model with and without the quota. Table 2.4 also provides strong evidence of one direction causality running from CDPRD to RGNP. This result has been previously investigated extensively by Hamilton (1983) regarding the macroeconomic effects of the oil price shocks on the U.S. economy. With respect to the inventory and production behavior, all variables except imports seem to Granger-cause CINV, while none of them appears to Granger-cause domestic production.

Tables 2.5 and 2.6 contain the results of two VAR systems concerning the refined petroleum products market. The series included in Table 2.5 are the refined petroleum products domestic real price (REFPRID), inventories (RINV), domestic output (ROUT) and imports (RIMP) respectively as well as the real GNP (RGNP). Table 2.6 includes the same variables except that RINV and ROUT have been replaced by the refined products sales (RSAL) and the real price of coal (PCOAL) respectively.

TABLE 2.5

Granger-Causality Tests with Specified Variables in the Refined Petroleum Products Market

Equation	REFPRID F	MS	RINV F	MS	ROUT F	MS	RIMP F	MS	RGNP F	MS
REFPRID	13.11	.0001	1.83	.18	3.84	.04	2.04	.15	1.58	.23
RINV	5.64	.01	1.40	.27	9.82	.0008	1.00	.35	4.26	.03
ROUT	.39	.68	.96	.39	10.65	.0005	1.33	.28	.55	.58
RIMP	3.15	.06	1.37	.27	.38	.68	12.0	.0002	.69	.51
RGNP	.73	.49	.29	.75	.13	.88	.20	.82	8.64	.001

Note: A constant and a linear trend were included in each equation.

TABLE 2.6

Granger-Causality Tests with specified variables in the Refined Petroleum Products Market

Equation	REFPRID F	MS	RINV F	MS	ROUT F	MS	RIMP F	MS	RGNP F	MS
REFPRID	4.62	.02	.79	.46	2.74	.08	3.89	.03	1.35	.28
RIMP	1.63	.22	13.70	.001	.65	.53	1.84	.18	.83	.45
RSAL	.82	.45	1.05	.37	9.34	.001	1.12	.34	1.98	.16
RGNP	7.28	.003	.35	.71	8.20	.002	38.56	.3E-07	15.31	.5E-04
PCOAL	2.64	.09	6.83	.004	1.73	.20	2.59	.10	29.92	.3E-06

Note: A constant and a linear trend were included in each equation.

As Table 2.5 indicates, refined products prices seem to be Granger-caused by lagged prices and domestic output, but none of the other variables (RINV, RIMP, RGNP) exert any causality on prices. Refined products imports (RIMP) seem to be Granger-caused by prices and their own lags. The fact that domestic production of refined petroleum products is strongly Granger-caused by its own lags is an indication of serially correlated productivity shocks and/or adjustment costs. In the system of Table 2.6, the price of coal has been introduced to test for any causality inferred from the price of refined products substitutes. As the results reveal, REFPRID is not Granger-caused by PCOAL. In conclusion, no clear evidence as to the direction of causality can be derived from the last two systems regarding the price behavior. Imports behavior, however, remains consistent with the earlier results.

Finally, a VAR system that is a mixture of crude and refined petroleum market variables was estimated. This system is described in Table 2.7 and includes the following series: the crude petroleum foreign real price (CFPRID), the refined petroleum products real price (REFPRID), the crude petroleum inventories (CINV), the refined petroleum products inventories (RINV), and the crude petroleum cumulative production (CUM).

TABLE 2.7

Granger-Causality Tests with Specified Variables in the Oil Industry

Equation	CFPRID F	MS	REFPRID F	MS	CINV F	MS	RINV F	MS	CUM F	MS
CFPRID	7.4	.003	11.02	.0004	.87	.43	.53	.59	2.92	.07
REFPRID	9.19	.001	16.68	.00003	.19	.83	3.10	.06	1.41	.26
CINV	1.53	.24	1.87	.18	2.95	.07	2.21	.13	.90	.42
RINV	2.58	.09	3.24	.06	7.61	.003	4.50	.02	.41	.67
CUM	.32	.73	.16	.86	.01	.98	.04	.96	287.45	0

Note: All variables were first-differenced and a constant was included in each equation.

The conclusions that derive from the results of Table 2.7 are that first, crude petroleum prices (CFPRID) appear to Granger-cause the refined products prices (REFPRID) and be Granger-caused by them (and similarly for the refined products prices), while cumulative production and refined inventories seem to weakly Granger-cause prices. Second, all variables are Granger-caused by their own lags except for the CINV series, while RINV is also Granger-caused by CINV and weakly by refined prices. Therefore, this system provides an indication about the endogeneity of the relevant variables which is crucial in justifying our initial claims that the American petroleum industry should be studied as a system of interrelated markets.

3. An LRE Model of the American Petroleum Industry

Following the observations of the previous section, the domestic demand for crude petroleum and the domestic supply of refined petroleum products are derived from a model that combines Kollintzas and Husted's (1984) LRE input inventory model with Blinder's (1982) LRE output inventory model. The domestic supply of crude petroleum is derived from Epple and Hansen's (1981) LRE exhaustible resource supply model. The domestic demand for refined petroleum products and the net foreign supplies of crude petroleum and refined petroleum products are modeled by traditional linear functions that are subject to additive stochastic disturbances.

The latter, the strategic petroleum reserve and the other government policy variables are represented by an AR law of motion that allows for intervention analysis of the type considered in Kollintzas and Geerts (1984). Figure 3.1 illustrates the model. The equilibrium of the model follows the rational expectations equilibrium formulation of Lucas and Prescott (1971) and Sargent (1979). That is, the representative domestic refinery and the representative domestic producer of crude petroleum consider the price of crude and refined petroleum products as being beyond their control, but their expectations about the path of these prices is such that the behavior of all domestic refineries and all domestic producers of crude petroleum products turns out to be consistent with the expectations of the representatives. To do this the representative domestic refinery and the representative domestic producer not only have to take into account each other's behavior but they also have to take into account the behavior of all other market participants. This behavior, of course, is summarized by the aggregate domestic demand for refined petroleum products, the aggregate foreign supplies of crude and refined petroleum products and the laws of motion of the domestic government's policy variables. It should be noted that the rational expectations equilibrium laws of motion induced by this equilibrium definition are for all practical purposes identical to those that would have been derived under what Hansen, Epple and Roberds (1985) call "time consistent equilibria". The class of time consistent equilibria includes the Nash equilibrium where agents recognize the effects of their actions on the aggregate constraints (15)-(16) but take each others' decisions as given. It also includes collusion equilibrium where the refinery and the crude petroleum producer maximize joint

Figure 3.1

The American Petroleum Industry as a System of Interrelated Markets

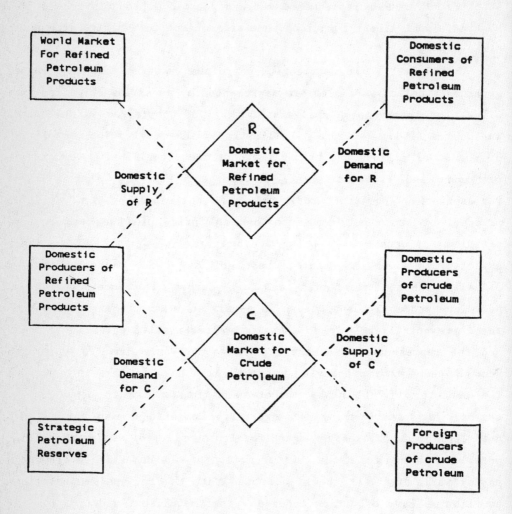

profits. "For all practical purposes," above, means that the equilibrium laws of motion associated with the other equilibria would have been observationally equivalent. Only the restrictions characterizing the coefficients of these laws would be different for each equilibrium.

3.1 The Representative Domestic Producer of Refined Petroleum Products

The domestic industry of refined petroleum products is assumed to consist of a fixed number of refineries, n_r, each of which produces R, using C and other factors. The production technology of the representative domestic producer of R is characterized by the fixed coefficient input requirements function:

$$C_{dc}(t) = \alpha R_{dp}(t), \quad \alpha > 0 \tag{1}$$

where $C_{dc}(t)$ and $R_{dp}(t)$ denote the representative domestic refiner's input of crude petroleum and output of refined petroleum products in period t, respectively.[15]

Further, it is assumed that the domestic refinery holds

---

[15] The hypothesis of the fixed coefficient input requirements function was tested in two ways. First, the OLS estimates of (1) (t-statistics in parentheses):

$$C_{dc}(t) = 32.35 + 1.04749\ R_{dp}(t) + u(t), \quad R^2 = .99, \quad SSE = 114184$$
$$\phantom{C_{dc}(t) = }(0.99)\quad (128.0)$$

indicate that the constant term is not statistically significantly different from zero and thus $C_{dc}(t)$ can be considered as a fixed ratio of $R_{dp}(t)$:

$$C_{dc}(t) = 1.05533\ R_{dp}(t) + u*(t)$$
$$\phantom{C_{dc}(t) = }(456.99)$$

Second, a Chow test of the hypothesis that α is statistically the same in the two periods T1 = 1947-1965 and T2 = 1966-1984 was performed. The SSR of the unrestricted case was found equal to 109340. Since the value of the F-statistic is

$$\frac{(114184. - 109340)/1}{109340./36} = 1.59 < F_{1,36} = 4.08,$$

the hypothesis is accepted.

inventories of both crude petroleum, C, and refined petroleum products, R, so that

$$C(t+1) = C(t) + C_{dd}(t) - \alpha R_{dp}(t) \tag{2}$$

$$R(t+1) = R(t) + R_{dp}(t) - R_{ds}(t)$$

where $C(t)$ and $R(t)$ are the inventories of C and R at the beginning of period t, respectively. $C_{dd}(t)$ and $R_{ds}(t)$ denote the amount of C purchased and the amount of R sold by the representative domestic refiner in period t, respectively.

At the beginning of any period τ, the representative domestic refiner chooses a contingency plan for its purchases of C, its production of R, and its sales of R, $\{C_{dd}(t), R_{dp}(t), R_{ds}(t)\}_{t=\tau}^{\infty}$, so as to maximize its expected discounted future stream of profits:

$$V^r[\{C_{dd}(t), R_{dp}(t), R_{ds}(t)\}_{t=\tau}^{\infty}, C(\tau), R(\tau); \tau]$$

$$= \lim_{T \to \infty} E_\tau^r \sum_{t=\tau}^{T} \beta_r^{t-\tau} \{[p_r(t) - q_r(t)]R_{ds}(t) - p_c(t)C_{dd}(t) - \Gamma[R_{dp}(t); t]$$

$$- \Delta[C(t), R(t); t]\} \tag{3}$$

subject to (1) and (2) and

$$C(\tau) = \tilde{C}$$

$$R(\tau) = \tilde{R} \tag{4}$$

where E is the mathematical expectations operator and $E_\tau^r(\cdot) = E(\cdot | \Omega_\tau^r)$ denotes that expectations are conditioned on the representative domestic refiner's information at the beginning of period τ. The content of the information set will be specified later; $\beta_r \varepsilon (0,1)$ is the discount factor; $p_r(t)$ is the domestic price of R, $q_r(t)$ is the excise tax on R, and $p_c(t)$ is the domestic price of C; finally, the functions

$\Gamma|^t$ and $\Delta|^t$ represent production costs and inventory holding costs respectively.

A quadratic specification of the cost functions $\Gamma|^t$ and $\Delta|^t$ is assumed, which is crucial for the certainty equivalence nature of the decision rules derived from linear rational expectations models. In particular, the following specification is adopted:

$$\Gamma|^t = c_1(t)R_{dp}(t) + \frac{1}{2} c_2 R_{dp}^2(t), \quad c_2 > 0 \tag{5}$$

$$\Delta|^t = \left[\beta_r^{-1}d_1(t-1) - d_1(t) + \frac{1}{2} d_2 C(t) + d_3 R(t)\right]C(t) +$$

$$\frac{1}{2} d_4 R(t)^2; \quad d_2, \ d_4, \ d_2 d_4 - d_3^2 \geq 0 \tag{6}$$

where $c_1(t)$ and $d_1(t)$ are time dependent parameters that incorporate the influence of random productivity shocks in the production and inventory holding processes of the representative domestic refiner. Technological progress may also affect these parameters. The restrictions on $\Gamma(\cdot)$ and $\Delta(\cdot)$ incorporate the usual convexity assumptions about production and inventory holding cost functions. The apparently complicated way $d_1(\cdot)$ enters in (6) is to facilitate the representation of the final solution. Thus we ignore adjustment costs in changing production levels and we ignore backlog costs. In the previous section we found evidence in justifying the second exclusion. However, we also found that lagged domestic production of refined products Granger causes current production. This can be explained by adjustment costs and/or by serially correlated productivity shocks. We leave adjustment costs for further research and we focus on serially correlated productivity shocks.

Finally, we follow Epple (1985) and model federal and state taxes by dividing all corporate federal, state, and local taxes payed by the industry with the industry's gross revenues to derive the effective "tax" per unit of output of the refined petroleum products industry. Of course, this approach is flawed in that it ignores the fact that firms respond differently to different taxes and in particular, that they respond to statutory rather than "effective" taxes. Thus, we effectively ignore incentive programs that may have increased capital spending and lowered production costs.

3.2 The Domestic Demand for Refined Petroleum Products

The domestic producers of refined products are faced with a downward sloping linear demand curve given by:

$$R_{dd}(t) = a_1(t)/a_2 - (1/a_2)p_r(t) + (a_3/a_2)p_s(t) + (a_4/a_2)Y(t), a_2 > 0 \quad (7)$$

where $R_{dd}(t)$ denotes the amount of R demanded by the domestic consumers, $a_1(t)$ is a time dependent parameter that incorporates the influence of random demand shocks, $p_s(t)$ is the price of domestic energy substitutes of refined petroleum products and $Y(t)$ is real domestic income. For linear rational expectations modeling purposes, this demand curve as well as the other demand (supply) curves that will be specified shortly are assumed to be linear.

3.3 The Net Foreign Supply of Refined Petroleum Products

The net foreign supply of refined petroleum products is specified as follows:

$$R_{fs}(t) = b_1(t) + b_2[p_r(t) - r_r(t)]; \quad b_2 > -\frac{1}{a_2} \quad (8)$$

where $R_{fs}(t)$, the amount of R supplied by foreigners in period t; $b_1(t)$ is a time dependent parameter that incorporates the influence of random demand (supply) shocks and $r_r(t)$ is the tariff (subsidy) on imported (exported) units of the refined product in period t.

3.4 The Representative Domestic Producer of Crude Petroleum

The domestic industry producing C consists of n_c firms. At the beginning of any period τ, the reprentative domestic producer of crude petroleum chooses a contingency plan for its production, $\{C_{ds}(t)\}_{t=\tau}^{\infty}$, so as to maximize its expected net worth:

$$V^c\left[\{C_{ds}(t)\}_{t=\tau}^{\infty}, S(\tau); \tau\right]$$

$$= \lim_{T \to \infty} E_\tau^c \sum_{t=\tau}^{T} \beta_c^{t-\tau} \left\{ \left[p_c(t) - q_c(t) \right] C_{ds}(t) - \theta \left[C_{ds}(t), S(t); t \right] \right\} \qquad (9)$$

subject to

$$S(t+1) = S(t) + C_{ds}(t) \qquad (10)$$

$$S(\tau) = \bar{S} \qquad (11)$$

where $\beta_c \in (0,1)$ is the discount factor of the representative domestic producer of C, $q_c(t)$ is the excise tax on C; $\theta|^t$ represents production costs in period t and $S(t)$ is the cumulative amount of crude petroleum that has been produced up to the beginning of period t. We model taxes in a manner similar to 3.1.

Again, for linear rational expectations purposes, it is assumed that the cost function $\theta|^t$ is quadratic and, in particular, that

$$\theta|^t = [e_1(t) + (1/2)e_2 C_{ds}(t) + e_3 S(t)] C_{ds}(t); \quad e_2, e_3 \geq 0 \qquad (12)$$

where $e_1(t)$ is a time dependent parameter that incorporates the influence of random productivity shocks. The reason for including $S(t)$ in $\theta|^t$ is to capture the exhaustible nature of C (i.e. $e_3 > 0$). The preceding model is essentially that of Epple and Hansen (1981). Following them, it will be necessary to further restrict the parameters of $\theta|^t$. Federal and state taxes are handled in a manner similar to Section 3.1.

3.5 The Foreign Producers of Crude Petroleum

The supply of the foreign producers of crude petroleum is given by:

$$C_{fs}(t) = \begin{cases} f_1(t)/f_2 + 1/f_2 \, [p_c(t) - r_c(t)], \; f_2 > 0, \text{ without quota} \\ \\ W(t) \hspace{5cm}, \text{ with quota} \end{cases} \qquad (13)$$

where $f_1(t)$ is a time dependent parameter that incorporates the influence of random supply shocks; $r_c(t)$ and $W(t)$ are the tariff and the quota on the imported units of crude petroleum in period t, respectively.[16]

3.6 The Government

As we have already specified, government taxes domestic producers of crude and refined petroleum products and imposes tariffs on imports of these products. Moreover, the government may impose a quota on imported crude petroleum products. In addition, the government may enter directly into the crude petroleum market by acquiring and disposing of "strategic petroleum reserves." We assume that the change in strategic petroleum reserves, $X(t)$, along with the other government policy variables follow an AR law of motion to be specified shortly. Our choice to model the change in strategic petroleum reserves by a fixed law of motion was not so much in order to avoid complexity but mostly because it is not clear how the government's behavior should be modeled.

3.7 Stochastic Environment, Information, and Expectations

Let

$$p(t) = [p_c(t), p_r(t)]'$$

$$q(t) = [q_c(t), q_r(t)]'$$

$$r(t) = [r_c(t), r_r(t)]'$$

$$s(t) = [a_1(t), b_1(t), c_1(t), d_1(t), e_1(t), f_1(t)]'$$

$$w(t) = [p_s(t), W(t), X(t), Y(t)]'$$

$$v(t) = [q(t)', r(t)', w(t)', s(t)']'$$

---

[16] Crude petroleum product tariffs were imposed during February 1975 to January 1976, while refined petroleum products tariffs were imposed during June to August 1975. Earlier in 1973, a system of import license fees was introduced in replacement of the Mandatory Oil Import Program. These tariffs and fees, however, were quite small with numerous exemptions and exceptions. See Kalt (1982, p. 8 and 115).

$$u(t) = [n_c C_{ds}(t), n_r R_{dp}(t), n_r R_{ds}(t)]'$$
$$x(t) = [n_r C(t), n_r R(t), n_c S(t)]'$$

The representative domestic producer of crude petroleum and the representative domestic producer of refined products are assumed to share the same stochastic environment, as represented by the process $\{v(\cdot)\}$ and the same information, as represented by the sequence of information sets $\{\Omega_t\}_{t=\tau}^{\infty}$, that is, $\Omega_t^c = \Omega_t^r = \Omega_t$. Moreover, the sequence of the information sets is strictly increasing:

$$\Omega_t \subset \Omega_{t+1}, \quad \forall t \in \{\tau, \tau+1, \ldots\}$$

and its elements include the history generated by the process $\{v(\cdot), p(\cdot), u(\cdot), x(\cdot)\}$. That is:

$$\{v(t'), p(t'), u(t'), x(t')\}_{t'=-\infty}^{t} \subset \Omega_t, \quad \forall t \in \{\tau, \tau+1, \ldots\}$$

Expectations are assumed to be rational. That is: (a) the subjective law of motion of the representative domestic producer of C and the representative domestic producer of R about $\{v(\cdot)\}$ and the objective law of motion of that process are identical, and (b) the subjective law of motion of these two producers about $\{p(\cdot), u(\cdot), x(\cdot)\}$ and the rational expectations equilibrium laws of motion (to be defined in the next subsection) of that process are identical.

Finally, it is assumed that the process $\{v(\cdot)\}$ is decomposed into a deterministic and a stochastic components as follows:

$$v(t) = v_c + v_\ell t + v_q t^2 + v_s(t) \tag{14}$$

The vector v_c contains the constant terms, v_ℓ and v_q are vectors of coefficients to the linear trend and quadratic trend respectively, and finally $v_s(t)$ is a stochastic vector which is assumed to follow an autoregressive law of motion:

$$A(L)v_s(t) = \varepsilon(t) \tag{14a}$$

where $v_s(t)$ and $\varepsilon(t)$ are each (nx1), and also

$$A(z) = A_0 + A_1 z - \ldots + A_n z^n \qquad (14b)$$

such that

$$\{z \varepsilon \mathbb{C} | \det A(z) = 0\} \cap \{z \varepsilon \mathbb{C} | |z| \leq \beta^{1/2}\} = \emptyset \qquad (14c)$$

The A_j's are square matrices comformable in dimension with the vector $v_s(t)$, L is the lag operator, and $\{\varepsilon(\cdot)\}$ is a vector white noise process. Condition (14b) means that the roots of the polynomial det $A(z)=0$ lie outside the circle of radius $\beta^{1/2}$, which ensures that $\{v_s(\cdot)\}$ is of mean geometric order less than $\beta^{-1/2}$. This is a necessary condition for the convergence of the infinite sums in (3) and (9). Finally, we allow for v_c to be a step like function in order to capture the interventions we alluded to in the previous section. Thus, for example, f_{1c} is allowed to "jump" in 73 and 78 so as to capture the effects of the Arab-Israeli War and the Iranian Crisis, respectively. Details can be found in the Data Appendix.

3.8 Equilibrium

Adapting the Lucas and Prescott (1971) and Sargent (1979) rational expectations equilibrium, we define a rational expectations equilibrium in the markets for C and R as a sequence $\{p^*(t), u^*(t), v^*(t)\}_{t=\tau}^{\infty}$ such that:

(a) Given $\{E_\tau p(t), v(t)\}_{t=\tau}^{\infty} = \{p^*(t), v^*(t)\}_{t=\tau}^{\infty}$, $\{C_{dd}^*(t), R_{dp}^*(t), R_{ds}^*(t)\}_{t=\tau}^{\infty}$, is a solution to the representative domestic refiner's problem.

(b) Given $\{E_\tau p(t), v(t)\}_{t=\tau}^{\infty} = \{p^*(t), v^*(t)\}_{t=\tau}^{\infty}$, $\{C_{ds}^*(t)\}_{t=\tau}^{\infty}$ is a solution to the representative domestic crude petroleum producer's problem.

(c) Given $\{E_\tau u(t), E_\tau v(t)\}_{t=\tau}^{\infty} = \{u^*(t), v^*(t)\}_{t=\tau}^{\infty}$, $\{p^*(t)\}_{t=\tau}^{\infty}$ clears the markets for C and R:

$$n_r R_{ds}^* + R_{fs}^*(t) = a_1(t)^*/a_2 - (1/a_2)p_r^*(t) - (a_3/a_2)p_s^*(t) - (a_4/a)Y^*(t) \quad (15)$$

$$R_{fs}^*(t) = b_1^*(t) + b_2[p_r^*(t) - r_r^*(t)] \quad (16)$$

$$n_c C_{ds}^*(t) + C_{fs}^*(t) = n_r C_{dd}^*(t) + X^*(t) \quad (17)$$

$$C_{fs}^*(t) \begin{cases} f_1(t)/f_2 + [p_c^*(t) - r_c^*(t)]/f_2, & \text{without quota} \\ \\ W^*(t) & , \text{with quota} \end{cases} \quad (18)$$

4. The Solution of the Model

4.1 The Forward Looking Solution

Proposition 1: $\{p^*(t), u^*(t), v^*(t)\}_{t=\tau}^{\infty}$ is a rational expectations equilibrium in the markets for C and R if and only if:

$$p^*(t) = Qu^*(t) + Pv^*(t) \quad (19)$$

$$n^*(t) = G^{-1} x^*(t+1) - G^{-1} x^*(t) + Hv^*(t) \quad (20)$$

where:

$$Q = \begin{bmatrix} 0 & -c_2/\alpha & -a_2/\alpha g \\ 0 & 0 & -a_2/g \end{bmatrix}$$

$$P = \begin{bmatrix} 0 & -1/\alpha & 0 & a_2 b_2/\alpha g & a_3/\alpha g & 0 & 0 & a_4/\alpha g & 1/\alpha g \\ 0 & 0 & 0 & a_2 b_2/g & a_3/g & 0 & 0 & a_4/g & 1/g \end{bmatrix}$$

$$\begin{bmatrix} -a_2/\alpha g & -1/\alpha & 0 & 0 & 0 \\ -a_2/g & 0 & 0 & 0 & 0 \end{bmatrix}, \quad g = (1 + a_2 b_2)$$

and its associated state path $\{x^*(t)\}_{t=\tau}^{\infty}$ satisfies the following:

$$JE_t x^*(t+2) - [(1+\beta^{-1})J+F]E_t x^*(t+1) + \beta^{-1} J E_t x^*(t) =$$

$$(B-JH)E_t \, [\beta^{-1}v^*(t)-v^*(t+1)] \tag{21}$$

$$(H,J) = \begin{cases} (H_0, J_0), & \text{without quota} \\ (H_1, J_1), & \text{with quota} \end{cases}$$

$$x(\tau) = \bar{x} = [n_r \bar{C} \quad n_r \bar{R} \quad n_c \bar{S}], \tag{22}$$

$$\beta^{(T-\tau)/2} E_\tau x(T) \to 0 \quad \text{as} \quad T \to \infty, \quad V\bar{x} \geq 0 \tag{23}$$

where:

$$B = \begin{bmatrix} 0 & -1/\alpha & 0 & a_2 b_2/\alpha g & a_3/\alpha g & 0 & 0 & a_4/\alpha g \\ 0 & -1 & 0 & a_2 b_2/g & a_3/g & 0 & 0 & a_4/g \\ 1 & 1/\alpha & 0 & -a_2 b_2/\alpha g & -a_3/\alpha g & 0 & 0 & -a_4/\alpha g \\ 1/\alpha g & -a_2/\alpha g & -1/\alpha & 1 & 0 & 0 & & \\ 1/g & -a_2/\alpha g & 0 & 0 & 0 & 0 & & \\ -1/\alpha g & a_2/\alpha g & 1/\alpha & 0 & 1 & 0 & & \end{bmatrix}$$

$$D = \begin{bmatrix} 0 & -n_r^{-1} c_2/\alpha & -a_2/\alpha g \\ 0 & 0 & -a_2/g \\ n_c^{-1}(e_2-e_3) & n_r^{-1} c_2/\alpha & a_2/\alpha g \end{bmatrix}$$

$$F = \begin{bmatrix} n_r^{-1} d_2 & n_r^{-1} d_3 & 0 \\ n_r^{-1} d_3 & n_r^{-1} d_4 & 0 \\ 0 & 0 & n_c^{-1}(\beta^{-1}-1)e_3 \end{bmatrix}$$

$$J_i = DG_i^{-1} \quad (i=0,1)$$

$$G_0 = \begin{pmatrix} 1 & -\alpha - (n_r^{-1} c_2/\alpha f_2) & -a_2/\alpha f_2 g \\ 0 & 1 & -1 \\ 1 & 0 & 0 \end{pmatrix}$$

$$G_1 = \begin{pmatrix} 1 & -\alpha & 0 \\ 0 & 1 & -1 \\ 1 & 0 & 0 \end{pmatrix}$$

$$H_0 = \begin{pmatrix} 0 & -1/\alpha f_2 & -1/f_2 & \dfrac{a_2 b_2}{\alpha f_2 g} & a_3/\alpha f_2 g & 0 & -1 & \dfrac{a_4}{\alpha f_2 g} \\ 0 & 0 & 0 & 0 & 0 & 0 & 0 & 0 \\ 0 & 0 & 0 & 0 & 0 & 0 & 0 & 0 \\ \dfrac{1}{\alpha f_2 g} & \dfrac{-a_2}{\alpha f_2 g} & \dfrac{-1}{\alpha f_2} & 0 & 0 & \dfrac{1}{f_2} & & \\ 0 & 0 & 0 & 0 & 0 & 0 & & \\ 0 & 0 & 0 & 0 & 0 & 0 & & \end{pmatrix}$$

$$H_1 = \begin{pmatrix} 0 & 0 & 0 & 0 & 0 & 1 & -1 & 0 & 0 & 0 & 0 & 0 & 0 & 0 \\ 0 & 0 & 0 & 0 & 0 & 0 & 0 & 0 & 0 & 0 & 0 & 0 & 0 & 0 \\ 0 & 0 & 0 & 0 & 0 & 0 & 0 & 0 & 0 & 0 & 0 & 0 & 0 & 0 \end{pmatrix}$$

Proof: In Appendix A.

Equation (19) follows from the market clearing conditions (15)-(18) and the fact that only any two of the components of $(C_{dd}(t), R_{dp}(t), R_{ds}(t))$ can be chosen independently of each other. The second order matrix stochastic difference equation (20) is a combination of the Euler conditions pertaining to the problems of the representative firms and the market clearing conditions. As usual, these Euler conditions characterize intertemporal cost/benefit tradeoffs. Thus, for example, the Euler equation pertaining to the representative refiner [see equation (A.9) in the Appendix] requires that the marginal cost of acquiring one unit of crude petroleum in any period must be equal to the expected discounted future stream of cost savings stemming from not having to acquire the same unit in the next period. Condition (23) is an asymptotic stability requirement. This condition is necessary because it implies a specific value (i.e. zero) for the "constant of integration" associated with the general solution to (21); while any other value for this constant would yield strictly smaller values for the objective functions in (3) and (9). It can be easily shown that given the globally concave nature of these functions, the necessary conditions (19)-(23) are also sufficient for a solution to the representative producer's problems. Hence, it follows by definition of a rational expectations equilibrium that (19)-(23) are necessary and sufficient for $\{u(t), p(t), v(t)\}_{t=\tau}^{\infty}$ to be a rational expectations equilibrium in the markets for C and R.

Lastly, it remains to solve (21) subject to (22) and (23). There are a number of ways for doing so. However, since J_i turns out to be symmetric and F is symmetric positive semidefinite, a more efficient method is employed which allows for all structural parameters (i.e. the elements of J_i, F, B, and H_i) to be identified uniquely as explained by the following proposition.

<u>Proposition 2</u>: Provided that $|J| \neq 0$ and $|x'[(1+\beta^{-1})J+F]x/x'Jx| > 2\beta^{-1/2}$ $\forall x \neq 0$, the unique solution to (21) subject to (22) and (23) is given by

$$x^*(t+1) = \Lambda x^*(t) + M(F)E_t[\beta^{-1}v(t)-v(t+1)] \qquad (24)$$

where F is the forward shift operator denoted by $F^j E_t x(t) = E_t x(t+j)$; Λ and $M(F)$ are defined as follows:

$$\Lambda = K\lambda K^{-1} = K \, \text{diag}\{\lambda_1,\lambda_2,\lambda_3\}K^{-1}$$

$$M(F) = \sum_{i=1}^{3} \kappa_i \kappa_i' \, [\lambda_i + (\beta\lambda_i)^{-1} - (1+\beta^{-1})]\beta\lambda_i \sum_{j=0}^{\infty} (\beta\lambda_i F)^j \, (JH-B)$$

where K denotes a non-singular matrix with columns κ_j, i.e., $K=[\kappa_j]$ (j=1,2,3), such that

$$K'JK = \varsigma = \text{diag } \{\varsigma_1, \varsigma_2, \varsigma_3\}$$

$$K'FK = I = \text{diag}\{1,1,1\}$$

and λ_j is the smallest modulus root of the characteristic equation:

$$z^2 - [(1+\beta^{-1})+\varsigma_j^{-1}]z + \beta^{-1} = 0 \qquad (j=1,2,3)$$

Proof: See Kollintzas (1985).

The restriction $|J| \neq 0$ is a regularity requirement. If J is singular it means that at least one component of $(n_r C(t), n_r R(t), n_c S(t))$ can be expressed in terms of its other components. In this case at least one decision rule is "myopic", and thus the problem can be formulated so that the ensuing lower dimension counterpart of J be nonsingular. However, for empirical purposes, the assumption $|J| \neq 0$ is inconsequential.

The restriction $|x'[(1+\beta^{-1})J+F]x/x'Jx| > 2\beta^{-1/2}$ is a stability requirement. Its effect is to make the λ_j's less than $\beta^{-1/2}$ in modulus (in which case the λ_j's are real). The latter implies that the sequence $\{x(\cdot)\}$ is of mean geometric order less than $\beta^{-1/2}$, and that the maximum values of the objective functionals in (3) and (9) are finite. Both conditions can be automatically satisfied given some additional restrictions on the structural parameters of the model. Thus, for example, $b_2 > 0$ (upward sloping net foreign supply); $d_2 > 0$, $d_4 > 0$, $(d_2 d_4 - d_3^2) > 0$ (strictly convex inventory holding costs); $e_3 > 0$ (C is an exhaustible resource); and $(e_2 - e_3) > 0$ (the Epple and Hansen restriction) imply both conditions. In fact, in this case J and F are positive definite and imply that the roots λ_j's are between zero and one. In other words, any shift in the exogenous variables will lead to smooth monotonic adjustments in the endogenous variables.

It should be finally emphasized that the elements of Λ and $M(\cdot)$, uniquely identify the structural parameters of the model. Equations (19), (24), and (14) express the linear rational expectations laws of motion in the markets for crude petroleum and refined petroleum products in terms of the infinite weighted sums of conditional expectations of the exogenous variables. Although, these equations are useful for comparative dynamics analysis, they are not useful for estimation.

It can be easily verified that:

$$M(F)E_t[\beta \tilde{v}(t) - v(t+1)] = \sum_{i=1}^{3} \tilde{N}_i v(t) + \sum_{i=1}^{3} \hat{N}_i \sum_{j=0}^{3} (\beta\lambda_i)^j E_t v(t+j) \quad (25)$$

where

$$\tilde{N}_i = [\lambda_i + \beta\lambda_i)^{-1} - (1+\beta^{-1})]\kappa_i \kappa_i'(JH-B) = \varsigma_i^{-1}\kappa_i \kappa_i'(JH-B)$$

$$\hat{N}_i = -(1-\lambda_i)[\lambda_i+(\beta\lambda_i)^{-1}-(1+\beta^{-1})\kappa_i \kappa_i'(JH-B) = -(1-\lambda_i)\varsigma_i^{-1}\kappa_i\kappa_i'(JH-B)$$

Thus, the equilibrium laws of motion (24) can be written as

$$(I-\Lambda L)x(t+1) = \sum_{i=1}^{3} [\tilde{N}_i v(t) + \hat{N}_i \sum_{j=0}^{\infty} (\beta\lambda_i)^j E_t v(t+j)] \quad (26)$$

To derive an estimable form of (26) the term $E_t v(t+j)$ must be eliminated by expressing it as functions of variables known to the agent at time t. This is pursued in the next section which explains the derivation of the backward looking solution of the previously obtained equilibrium laws of motion.

4.2 The Backward Looking Solution

The linear rational expectations equilibrium laws of motion in the markets for C and R are defined by (14)-(14b), (19) and (26). In order to express them in an estimable form one must first, transform the infinite leads in (26) in terms of the elements in the information set Ω; and second, to substitute the "observed by the econometrician" elements of Ω_t for the unobserved components included in the process $\{v(\cdot)\}$. Both of these tasks can be carried out by following results in Hansen and Sargent (1980) as explained below.

First, in view of assumption (14) and after collecting terms, the equilibrium laws of motion (26) give the following:

$$(I-\Lambda L)x(t+1) = \sum_{i=1}^{3}\{[N_i+\tilde{N}_i(\hat{1}-\beta\lambda_i)]v_c^{-1}+N_i\beta\hat{\lambda}_i(1-\beta\lambda_i)\ v_\ell^{-2}$$

$$+\hat{N}_i\beta\lambda_i(1+\beta\lambda_i)(1-\beta\lambda_i)^{-3}v_q\} + \sum_{i=1}^{3}\{\tilde{N}_i+\hat{N}_i(1-\beta\lambda_1)^{-1}\}v_\ell$$

$$+2\hat{N}_i\beta\lambda_i(1-\beta\lambda_i)^{-2}v_q\}t + \sum_{i=1}^{3}[\tilde{N}_i+\hat{N}_i(1-\beta\lambda_i)^{-1}]v_q t^2$$

$$+ \sum_{i=1}^{3}\tilde{N}_i v_s(t) + \sum_{i=1}^{3}\hat{N}_i \sum_{j=0}^{\infty}(\beta\lambda_i)^j E_t v_s(t+j) \qquad (27)$$

Then since $|\lambda_i| < \beta^{-1/2}$ for all $i=1,2,3$, and given assumption (14c) about the roots of $|A(z)| = 0$, it follows by Hansen and Sargent (1980, p. 99), that

$$\sum_{j=0}^{\infty}(\beta\lambda_i)^j E_t v_s(t+j) = A(\beta\lambda_i)^{-1}[A_0 + \sum_{j=1}^{n-1}\sum_{k=j+1}^{n}(\beta\lambda_i)^{k-j}A_k L^j]v_s(t) \qquad (28)$$

$$= T_i(L)v_s(t)$$

Substituting (28) in (27) and obvious notation we obtain

$$(I-\Lambda L)x(t+1) = U_0 + U_1 t + U_2 t^2 + U_s(L)v_s(t) \qquad (29)$$

where the U's are appropriately dimensioned and incorporate the following: U_0 is the matrix of the constant terms, U_1 and U_2 are the matrices of coefficients to the linear and quadratic trend respectively, and finally, the polynomial $U_s(L)$ given by

$$U_s(L) = \sum_{i=1}^{3}[\tilde{N}_i+\hat{N}_i T_i(L)] \qquad (30)$$

Equation (29) is the backward looking version of (24), since it is expressed in terms of variables known to the economic agents at time t. However, (29) cannot be estimated directly since the vector of shocks

$s(t)' = [a_1(t), b_1(t), c_1(t), d_1(t), e_1(t)]$ included in the process $\{v(\cdot)\}$ is not observed by the econometrician. As it turns out, this unobserved component of $\{v(\cdot)\}$ will give rise to disturbance terms in the regression equations derived below.

Thus, partitioning $U_s(L)$ and $v_s(t)$ in (29) according to (14a), we obtain

$$(I-\Lambda L)x(t+1) = U_0 + U_1 t + U_2 t^2 + U_q(L)q(t) + U_r(L)r(t) \qquad (31)$$

$$+ V_w(L) w_s(t) - U_{ss}(L) s_s(t)$$

Similarly, in view of (A.25) and given assumptions (14)-(14a), the equilibrium price equation (19) becomes

$$p(t) - QG^{-1}(I-L)x(t+1) \qquad (32)$$

$$= [QG^{-1}H - P]v(t) = -\theta_0 - \theta_1 t - \theta_2 t^2 - \theta_q q_s(t) - \theta_r r_s(t) - \theta_w w_s(t) - \theta_{ss} s_s(t)$$

where $\theta = QG^{-1}H - P$ and the superscripts denote the appropriate partitioning of θ according to $v(t)$ in (14)-(14a).

Finally, (14b) needs also partitioning according to (14a) so that

$$\begin{pmatrix} A_{qq} & A_{qr} & A_{qw} & A_{qs} \\ A_{rq} & A_{rr} & A_{rw} & A_{rs} \\ A_{wq} & A_{wr} & A_{ww} & A_{ws} \\ A_{sq} & A_{sr} & A_{sw} & A_{ss} \end{pmatrix} \begin{pmatrix} q_s(t) \\ r_s(t) \\ w_s(t) \\ s_s(t) \end{pmatrix} = \begin{pmatrix} \varepsilon_q(t) \\ \varepsilon_r(t) \\ \varepsilon_w(t) \\ \varepsilon_s(t) \end{pmatrix} \qquad (33)$$

where the A's denote appropriately dimensioned submatrices of the lag polynomial $A(L)$ in (14b). The lag operators have been dropped from the A's for purposes of parsimony.

Summarizing, we have shown that the backward looking linear rational expectations laws of motion are given by (31), (32) and (33).

It remains to eliminate the unobserved component s(t) from the system of these equations. Thus, from the third equation in (33) we have

$$s_s(t) = -A_{ss}^{-1}A_{sq}q(t) - A_{ss}^{-1}A_{sr}r_s(t) - A_{ss}^{-1}A_{sw}w_s(t) + A_{ss}^{-1}\varepsilon_s(t) \tag{34}$$

Substituting (34) back in the system of equations (31)-(33) and rearranging terms, we obtain the final form of the system described by (35) below.

As stated previously, the fact that the vector $s_s(t)$ is unobserved to the econometrician introduces the disturbance term $\varepsilon_i(t)$, $i=q,r,s,w$, in the equations of system (35) which can now be estimated empirically. This system has the form of an ARMAX model involving the processes $p(t)$, $x(t+1)$, $q_s(t)$, $r_s(t)$ and $w_s(t)$. In particular, it includes two price equations, one for crude petroleum prices, $p_c(t)$ and the other for the refined petroleum products prices, $p_r(t)$; three equations in $x(t+1)$ involving the inventories equations for crude, $C(t)$ and refined products, $R(t)$, and the cumulative production equation of crude petroleum, $S(t)$; four equations involving the processes of taxes/subsidies of crude, $q_{sc}(t)$ and refined $q_{sr}(t)$, and also the tariffs of crude $r_{sc}(t)$ and refined $r_{sr}(t)$, respectively; finally, the equations involving the vector of the exogenous variables $w_s(t)$.

The specification implied by the derived system has the great advantage over previous specifications in that it distinguishes between structural and expectations parameters as will be clear in the next chapter. Although the system is linear in the variables, their coefficients are nonlinear functions of the underlying parameters of the model. The discussion of the econometric issues that arise in estimating the above system is undertaken in the next section.

5. Empirical Results

Before proceeding with the details of the estimation process, a summary of the parameters to be estimated in the present model is given. These parameters fall into two categories: structural parameters and expectations parameters. The structural parameters are those that enter the agents' objective functions and the associated transition equations. These parameters are described in Table 5.1.

The expectation parameters characterize the laws of motion of the stochastic environment of the agents' problems and consist of a deterministic component and a stochastic component as shown in Table 5.2.

$$\begin{bmatrix} I & -QG^{-1}(I-L) & \theta_q - \theta_{ss} A_{ss}(L)^{-1} A_{sq}(L) & \theta_r - \theta_{ss} A_{ss}(L)^{-1} A_{sr}(L) \\ 0 & (1-\Lambda L) & U_{ss}(L) A_{ss}(L)^{-1} A_{sq}(L) - U_q(L) & U_{ss}(L) A_{ss}(L)^{-1} A_{sr}(L) - U_r(L) \\ 0 & 0 & A_{qq}(L) - A_{qs}(L) A_{ss}(L)^{-1} A_{sq}(L) & A_{qr}(L) - A_{qs}(L) A_{ss}(L)^{-1} A_{sr}(L) \\ 0 & 0 & A_{rq}(L) - A_{rs}(L) A_{ss}(L)^{-1} A_{sq}(L) & A_{rr}(L) - A_{rs}(L) A_{ss}(L)^{-1} A_{sr}(L) \\ 0 & 0 & A_{wq}(L) - A_{ws}(L) A_{ss}(L)^{-1} A_{sq}(L) & A_{wr}(L) - A_{ws}(L) A_{ss}(L)^{-1} A_{sr}(L) \end{bmatrix}$$

$$\begin{bmatrix} \theta_w - \theta_{ss} A_{ss}(L)^{-1} A_{sw}(L) \\ U_{ss}(L) A_{ss}(L)^{-1} A_{sw}(L) - U_w(L) \\ A_{qw}(L) - A_{qs}(L) A_{ss}(L)^{-1} A_{sw}(L) \\ A_{rw}(L) - A_{rs}(L) A_{ss}(L)^{-1} A_{sw}(L) \\ A_{ww}(L) - A_{ws}(L) A_{ss}(L)^{-1} A_{sw}(L) \end{bmatrix} \begin{bmatrix} p(t) \\ x(t+1) \\ q_s(t) \\ r_s(t) \\ w_s(t) \end{bmatrix} = \begin{bmatrix} -\theta_0 & -\theta_1 & -\theta_2 \\ U_0 & U_1 & U_2 \\ 0 & 0 & 0 \\ 0 & 0 & 0 \\ 0 & 0 & 0 \end{bmatrix}$$

$$\begin{bmatrix} 1 \\ t \\ t^2 \end{bmatrix} + \begin{bmatrix} -\theta_{ss} A_{ss}(L)^{-1} & 0 & 0 & 0 \\ U_{ss}(L) A_{ss}(L)^{-1} & 0 & 0 & 0 \\ -A_{qs}(L) A_{ss}(L)^{-1} & 0 & 0 & 0 \\ -A_{rs}(L) A_{ss}(L)^{-1} & 0 & 0 & 0 \\ -A_{ws}(L) A_{ss}(L)^{-1} & 0 & 0 & 0 \end{bmatrix} \begin{bmatrix} \varepsilon_s(t) \\ \varepsilon_q(t) \\ \varepsilon_r(t) \\ \varepsilon_w(t) \end{bmatrix} \quad (35)$$

TABLE 5.1: <u>Parameters of the Model: Structural Parameters</u>

β: The discount factor.

α: The fixed input/output ratio of the representative domestic refinery.

a_2: The negative of the inverse of the slope ($-1/a_2$) of the inverse domestic demand for refined petroleum products.

a_3: The coefficients of the real price of refined petroleum product substitutes in the inverse domestic demand for refined petroleum products.

a_4: The coefficient of real income in the inverse domestic demand for refined petroleum products.

b_2: The slope of the net foreign supply of refined petroleum products.

c_2: The slope of the marginal production cost of the representative domestic refiner.

d_2: The slope of the marginal inventory holding cost of the representative domestic refiner with respect to crude petroleum inventories.

d_3: The coefficient of the cross product of crude petroleum and refined products inventories in the inventory holding cost function.

d_4: The slope of the marginal inventory holding cost of the representative domestic refiner with respect to refined petroleum product inventories.

e_2: The slope of the marginal production cost of the representative domestic producer of crude petroleum.

e_3: The coefficient of the cross product of current and cumulative crude petroleum production in the production function of the representative domestic producer of crude petroleum.

f_2: The inverse of the slope ($1/f_2$) of the net foreign supply of crude petroleum.

TABLE 5.2: Parameters of the Model: Expectations Parameters

Parameters characterizing the deterministic component $(v_c + v_\ell t + v_q t^2)$ of the exogenous state variables.

a_{1c}: The constant in the intercept of the inverse domestic demand for refined petroleum products.

b_{1c}: The constant in the intercept of the net foreign supply of refined petroleum products.

c_{1c}: The constant in the intercept of the marginal cost of production of the representative domestic refiner.

d_{1c}: The constant in the intercept of the marginal inventory holding cost of the representative domestic refiner with respect to crude petroleum inventories.

e_{1c}: The constant in the intercept of the marginal cost of production of the representative domestic crude petroleum producer.

f_{1c}: The negative of the constant of the intercept of the inverse net foreign supply of crude petroleum products.

q_{cc}: The constant in the law of motion of the "excise tax" on crude petroleum products.

q_{rc}: The constant in the law of motion of the "excise tax" on refined petroleum products.

r_{cc}: The constant in the law of motion of the tariff on imported crude petroleum products.

r_{rc}: The constant in the law of motion of the tariff on imported refined petroleum products.

w_{sc}: The constant in the law of motion of the real price of refined petroleum product substitutes.

w_{wc}: The constant in the law of motion of the quota on crude petroleum products.

w_{xc}: The constant in the law of motion of the change in the Strategic Petroleum Reserve.

w_{yc}: The constant in the law of motion of domestic real income.

$a_{1\ell}$: The coefficient of the linear trend term in the intercept of the inverse domestic demand for refined petroleum products.

a_{1q}: The coefficient of the quadratic trend term in the intercept of the inverse domestic demand for refined petroleum products.

Likewise for $b_{1\ell}, c_{1\ell}, \ldots, w_{y\ell}, b_{1q}, c_{1q}, \ldots, w_{yq}$

Parameters characterizing the stochastic component ($v_s(t)$) of the exogenous state variables.

A_i (i=1,...,n): The coefficients of the $A(z)$ matrix polynomial in the law of motion of the stochastic component of the exogenous state variables.

Given the relatively small data sample (38 observations) the simultaneous estimation of all the model parameters described in Tables 5.1 and 5.2 is not possible. This being the case, some of the less crucial parameters were restricted a priori, while others were estimated separately. First, specific assumptions were imposed on the form of the matrix polynomial $A(z)$. Thus it was required that

$$A(z) = I - Az \qquad (36)$$

where

$$A = \text{diag}\{A_{qq}L, A_{rr}L, A_{ww}L, A_{ss}L\}$$

and the A_{ii}'s (i=q,r,s,w) are appropriately dimensioned matrices. The zero restrictions above imply that any given element of the $\{q_s(\cdot), r_s(\cdot), w_s(\cdot), s_s(\cdot)\}$ process fails to Granger cause any other element of that process. The AR(1) specification was required to keep the length of lags in the final form to a minimum and, therefore, reserve the desired degrees of freedom for statistical inference.

Under these assumptions, the theoretical model (35) reduces to

$$\begin{pmatrix} I & -QG^{-1}(1-L) & \theta_q & \theta_r & \theta_w \\ 0 & (I-\Lambda L) & -U_q(L) & -U_r(L) & -U_w(L) \\ 0 & 0 & (I-A_{qq}(L)) & 0 & 0 \\ 0 & 0 & 0 & (I-A_{rr}L) & 0 \\ 0 & 0 & 0 & 0 & (I-A_{ww}L) \end{pmatrix} \begin{pmatrix} p(t) \\ x(t+1) \\ q_s(t) \\ r_s(t) \\ w_s(t) \end{pmatrix} =$$

$$\begin{pmatrix} -\theta_0 & -\theta_1 & -\theta_2 \\ U_0 & U_1 & U_2 \\ 0 & 0 & 0 \\ 0 & 0 & 0 \\ 0 & 0 & 0 \end{pmatrix} \begin{pmatrix} 1 \\ t \\ t^2 \end{pmatrix} + \begin{pmatrix} -\theta_s(I-A_{ss}L)^{-1} & 0 & 0 & 0 \\ U_s(I-A_{ss}L)^{-1} & 0 & 0 & 0 \\ 0 & I & 0 & 0 \\ 0 & 0 & I & 0 \\ 0 & 0 & 0 & I \end{pmatrix} \begin{pmatrix} \varepsilon_s \\ \varepsilon_q \\ \varepsilon_r \\ \varepsilon_w \end{pmatrix} \quad (37)$$

It should be observed that system (37) takes the form of a block-recursive system and, therefore, can be estimated separately. Then, under the previous simplifications, the upper part of this system, consisting of the first five rows, can be written as

$$(I-A_{ss}L)B^{-1} \begin{pmatrix} I & QG^{-1}(I-L) & \theta_q & \theta_r & \theta_w \\ 0 & (I-\Lambda L) & -U_q(L) & -U_r(L) & -U_w(L) \end{pmatrix} \begin{pmatrix} p(t) \\ x(t+1) \\ q_s(t) \\ r_s(t) \\ w_s(t) \end{pmatrix}$$

$$= (I-A_{ss}L)B^{-1} \begin{pmatrix} -\theta_0 & -\theta_1 & -\theta_2 \\ U_0 & U_1 & U_2 \end{pmatrix} \begin{pmatrix} 1 \\ t \\ t^2 \end{pmatrix} + \varepsilon_s(t) \quad (38)$$

where

$$B = \begin{pmatrix} -\theta_s \\ U_s \end{pmatrix}$$

More compactly, this system has the ARMAX form:

$$\Phi_0 \Psi(t) + \Phi_1 \Psi(t-1) + \Phi_2 \Psi(t-2) = \Theta_0 \omega(t) + \varepsilon_s(t) \quad (39)$$

where $\Psi(t) = [p(t)'x(t+1)'q_s(t)'r_s(t)'w_s(t)']$; $\omega(t) = [1,t,t^2]'$, and the matrix coefficients Φ_i, $i=0,1,2$ and Θ_0 are appropriately dimensioned. It should be recalled that the elements of these matrices are non-linear functions of the structural parameters to be estimated. Moreover, θ_0 is time dependent. This is a consequence of the fact that v_c is time dependent in order to capture the intervention.

The concentracted likelihood function of system (39) is given by

$$L^* = -(TN/2)(\ln 2\pi + 1) - (T/2)\ln|\Sigma| + T\ln|\det\theta_{01}| \qquad (40)$$

where

$$\Sigma = T^{-1} \sum_{t=1}^{T} \varepsilon_s(t) \varepsilon_s(t)'$$

$$\varepsilon_s(t) = \Phi_0 \Psi(t) + \Phi_1 \Psi(t-1) + \Phi_2 \Psi(t-2) - \theta_0 \omega(t)$$

L^* denotes the logarithm of the concentrated likelihood function; T and N are the number of observations and the number of endogenous variables respectively; Σ is the variance-covariance matrix of the vector of residuals $\varepsilon_s(t)$; θ_{01} is the matrix of coefficients of the current endogenous variables included in the vector $\Psi(t)$, that is the subvectors $p(t)$ and $x(t+1)$.

Table 5.3 contains the ML estimates of the structural and expectations formation parameters. The standard errors reported in this table are the asymptotic standard errors obtained from the inverse of the Hessian of the likelihood function evaluated at the terminal point. All reported results were obtained by setting $\beta=.95$. This corresponds to an annual discount rate $(\beta^{-1}-1)$ of approximately 5%. The estimate of the input/output ratio $\alpha = 1.055$ was obtained from (1) by an OLS regression (see footnote 15).

It should be noted that some additional restrictions were imposed on the values of structural parameters in order to facilitate the estimation of the remaining terms. Thus, d_3 was set equal to zero, i.e., no interaction in the costs of holding inventories was assumed between crude petroleum and refined products inventories. However, there is no a priori theoretical restriction on the sign of this parameter. Moreover, following the indications of preliminary results, d_2 was set equal to d_4 (i.e., same slopes of the marginal inventory

TABLE 5.3: Parameter Estimates

Parameter	Final Estimate	Standard Error
a_2	5.11	.99
a_4	0.08	0.02
b_2	.62	.17
c_2	477.26	.99
$d_2 = d_4$	238.78	.99
e_2	8.48	.94
e_3	.56	.11
f_2	1.87	.09
a_{1c}	3.95	.98
c_{1c}	119.99	1.03
d_{1c}	-118.93	1.05
e_{1c}	-48.50	.99
f_{1c}	-1.88	.30
a_{11}	2.78	.96
c_{11}	-18.01	1.00
d_{11}	-31.02	.99
e_{11}	-3.76	.89
f_{11}	.14	.08
a_{s1}	1.003	.01
a_{s2}	1.035	.02
a_{s3}	.96	.01
a_{s4}	1.02	.03
a_{s5}	-.98	.01

Fixed Parameters: $\beta = .95$, $b_{1c} = 0$, $\alpha = 1.055$, $d_3 = 0$, $a_3 = 0$

holding cost of the representative domestic refiner with respect to crude petroleum and refined petroleum products inventories were assumed in the inventory holding cost function). Finally, only estimates of the linear trend coefficients are reported. The results using in addition a quadratic term were somewhat poorer and are not reported here.

Before proceeding with a critical evaluation of the empirical findings, it should be mentioned that an intervention analysis was used to capture the oil price shocks in the time series data of the model. As noted by Husted and Kollintzas (1982, 1984, 1987) in their empirical works on commodity markets, the presence of price shocks in the data introduces considerable difficulties in representing prices by ARMA models which can be alleviated by removing them through an intervention analysis. That is, shocks were captured by allowing for the intercepts in the price equations to be time-dependent. Thus, an appropriate step-like change in the mean of $f_1(.)$ was introduced to account for the two price shocks (the first associated with the 1973 Arab-Israeli War and the second one associated with the 1978-79 Iranian crisis, which led to dramatic price increases. Similar treatment was applied to the government policy variables.

At this point it is important to check the plausibility of the reported estimates in terms of the expected signs and magnitudes. As shown in Table 5.3, all parameter estimates are highly significant and conform to the theoretical restrictions.

In particular, the parameters a_2, c_2, d_2, e_2 and e_3 are positive as required by theory. The latter ($e_3 > 0$) provides evidence for the exhaustible nature of crude petroleum which confirms Epple's (1985) result. In addition, $e_2 - e_3 > 0$, which along with the previous conditions guarantee that the characteristic roots λ_i's are all between zero and one. Recall that the λ_i's (i=1,2,3) are the smallest absolute value roots of the characteristic equation associated with the law of motion of the process $\{x(\cdot)\}$ obtained by solving the optimization problems of both the representative domestic producers of crude and refined petroleum products. The values of λ_i's associated with the reported results are

$$\lambda_1 = 0.07 \quad \lambda_2 = 0.05 \quad \lambda_3 = 0.96$$

The fact that the λ_i's are positive implies a smooth monotonic adjustment in the inventories of crude petroleum and refined products as well as to the cumulative production of crude in response to a change in the exogenous variables over time. It is interesting to notice that these values of λ_i's imply very high rates of adjustments in the stocks of crude and refined petroleum products held by the representatitive refiner, but a relatively low speed of adjustment in the cumulative production by the representative producer of crude petroleum. This can be seen by writing (26) as

$$x(t+1) - x(t) = (I-\Lambda)[x^*(t)-x(t)] \tag{41}$$

where

$$x^*(t) = (I-\Lambda)^{-1} \sum_{i=1}^{3} \tilde{N}_i v(t) + \hat{N}_i \sum_{j=0}^{\infty} (\beta\lambda_i)^j E_t v(t+j)$$

and $x^*(t)$ is the target or desired level of $x(t)$. Equation (41) is the partial adjustment inventory model developed by Lovell (1961). The diagonal elements of Λ do not differ substantially from the values of λ_i's and hence $1-\lambda_i$ for all i, can be interpreted as the coefficient of adjustment in response to any anticipated change in the target level $x^*(t)$. Obviously, these coefficients $(1-\lambda_i)$ are very close to unity for the inventories implying an almost complete adjustment (93% or more) over a year. Conversely, only a very small portion of adjustment (less than 4%) is completed within a year for the cumulative production of crude petroleum. The high rates of adjustment give support for the production smoothing inventory model which has raised so many questions as a result of its failure to be verified empirically. What is more, this result conforms to the stylized facts described in Section 2, where it was found that the variance of output is actually less than the variance of sales in crude and refined petroleum products. A final point worth mentioning on the same issue is to see what happens with the values of the curvature parameters of the cost functions since they provide a motive for (or against) production smoothing. As Table 5.3 reveals, the estimated values of c_2 and d_2 (equal to d_4) are quite high. Hence, the steeply increasing marginal cost of holding inventories implied by d_2 (and d_4) should discourage refiners from using inventory movements to smooth production. However, the much higher rate at which the production marginal cost (measured by c_2) increases is expected to create a stronger motive for the refiner to smooth production.

Turning now to the remaining parameter estimates, first the importance of f_2 is noted, the inverse of which is the slope of the net foreign supply of imports of crude petroleum. This parameter has a statistically significant and positive value suggesting that the U.S. has some market power in the world market for crude petroleum. The elasticity of imports supply evaluated at the mean of the data is 5.16. Comparing this value with previous results, the only comparable evidence seems to be the result reported by Husted and Kollintzas (1987). Their estimate of the import supply elasticity is smaller than the present estimate. In their model the domestic refinery minimizes costs, subject to a sales constraint and imports are subject to adjustment costs. It is possible that our exclusion of adjustment costs has biased upward the estimate of f_2.

Especially interesting in an empirical analysis of this market is the price elasticity of refined products demand. In order to derive an estimate of this elasticity, the value of a_2 is used, the inverse of which is the slope of the domestic demand for refined products. Hence, the implied short-run elasticity of this demand evaluated at the mean of the data gives a value of .71. Since there is considerable comparative evidence, this result is compared with the results of previous studies shown in Table 5.4. Only one study (Alt, Bopp and Lady, 1976) has reported a value for all the refined products as a composite. Although the rest of the studies refer to gasoline demand elasticity, it is worth comparing these elasticities since gasoline carries the heavier weight among the other refined products in the sales. The evidence that the price elasticity of refined petroleum products is rather high is consistent with the belief that in response to the prices shocks of the seventies demand decreased fairly rapidly.

Finally, Table 5.5 presents a summary of some diagnostic statistics for each of the equations in the system. In general, the equations individually exhibit a good fit to the data as measured by the standard errors, which are quite small relative to the sample means. The reported R^2 also provide some notion of the overall explanatory power of the model. In all the equations these values suggest that a considerable proportion of the variance of the endogenous variable is explained by the model.

In order to provide some evidence as to the white noise properties of the residuals from the estimated model, the values of the Q statistic introduced by Box and Pierce can be used as a diagnostic checking of the estimated residuals. This statistic is defined as

$$Q_k = T \sum_{j=1}^{K} r_j^2 \qquad (42)$$

where r_j is the jth sample autocorrelation in the residuals. K is the maximum lag length in the autocorrelations and T is the sample size. The statistic is approximately distributed as $X^2(K-p-q)$, where p and q are the appropriate orders of an ARMA (p,q) process. However, in the case of an ARMAX model there is some ambiguity over the correct number of degrees of freedom. Here p is taken to be the maximum lag length in the right-hand side variables which is 2, and 9 is zero. Thus, the values of the Q statistic at the 20th lag (K=20) are reported in the last column of Table 5.4. Given the table value of the X^2 with K-p =18 degrees of freedom equal to 34.81 at the 0.01 critical level, the hypothesis of white noise is accepted for all but the $p_r(t)$ equation which is close to the rejection point and the inventory equation C(t) which is decisively rejected. There is syggestive evidence, therefore, of serial correlation in a couple of the model's equations. Although it is known what a serious impact autocorrelation can have on the

TABLE 5.4: Refined Petroleum Products Price Elasticities

Author	Short-run Price Elasticity	
	Weighted Average of Products	Gasoline
Present Model*	-0.71	-
Alt, Bopp & Lady (1976)**	-0.188	-0.30
Rice and Smith (1977)	-	-0.35
Cato, Rodekohr & Sweeney (1976)	-	-0.24
Ramsey, Rasche & Allen (1975)	-	-0.65
Kennedy (1974)	-	-0.47
Houthakker, Verleger & Sheehan (1974)	-	-0.43

*It refers to a composite of only four products: gasoline, kerosene, distillate and residual oils.
**It refers to all refined products.

TABLE 5.5: Summary Statistics

Equation	Mean of Dependent Variable	Standard Error of Estimation	R^2	q(20) Statistic
$p_c(t)$	6.428	.720	0.99	17.6
$p_r(t)$	13.939	1.560	0.98	33.9
$C(t)$	0.395	0.033	0.99	38.2
$R(t)$	0.537	0.049	0.99	16.8
$S(t)$	94.757	0.135	0.99	9.16

Note: The variables $C(t)$, $R(t)$, and $S(t)$ were measured in billions of barrels, while $p_c(t)$ and $p_r(t)$ in dollars per barrel.

sampling distribution of the estimates, unfortunately there is no "fix" available for this problem.[17]

Finally, an attempt was made to test the nonlinear and cross equation restrictions implied by the model. Since the model is not nested within a more general structure, it is not clear what test would be appropriate to apply in this case. A test that has been used in similar applications is the likelihood ratio test. This test statistic is constructed as minus two times the difference between the values of the log likelihood functions of the restricted and unrestricted systems, i.e., $-2(L^R-L^U)$. The statistic is asymptotically distributed as X^2_r, where r is the number of restrictions in the model. There is some ambiguity here as to what the relevant unrestricted model should be. In this case the unrestricted model is taken to be the system of linear equations for p_c, p_r, C, R, and S with the same right-hand side variables as the theoretical model but with no non-linear or cross equation restrictions on the coefficient estimates except for the zero restrictions. That is, a system equivalent to (39) was estimated with the exclusion of those variables that enter with zero coefficients in the theoretical model. The number of restrictions is taken to be the number of freely estimated parameters in the unrestricted model minus the number of parameters estimated in the restricted system. Thus, the appropriate likelihood-ratio statistic for this model is distributed as X^2 with r=35 degrees of freedom. The table value for $X^2(35)$ at the 0.01 critical level is 57.29. The value of the likelihood-ratio test computed from the data sample was found equal to 410.46. Under the null hypothesis that the restricted model is consistent with the sample data, this value indicated decisively the rejection of the hypothesis at the 1% or higher level of significance. The fact that the theoretical model failed this test is hardly surprising given the extremely complicated set of restrictions imposed by the theory.[18]

6. Conclusion

---

[17] See Goldfeld and Quandt (1972) for the autocorrelation problem in simultaneous equation systems in general and, in particular, Husted and Kollintzas (1984) for a discussion of this problem in LRE models.

[18] There is a number of questions that are open for discussion on this issue. See Husted and Kollintzas (1987).

An "integrated" model of the American petroleum industry using the linear rational expectations methodology was developed and estimated in this study. The model takes into account the storable and exhaustible nature of petroleum as well as the dynamic and strategic interaction of agents operating within the stochastic environment of the two mutually interrelated markets of crude petroleum and refined petroleum products. The model was built on two optimization problems: (a) the problem of the representative domestic producer of refined petroleum products who was assumed to maximize the expected present value of revenues subject to production and inventory holding costs; and (b) the problem of the representative producer of crude petroleum who was assumed to solve a similar problem subject to production costs that account for the exhaustible nature of petroleum. Furthermore, the domestic demand and foreign supply of refined petroleum products and the foreign supply of crude petroleum were modeled by traditional linear functions that are subject to additive stochastic disturbances. Although these last relationships were not derived explicitly from an optimization process, considerable structure was imposed on their stochastic terms.

In Section 3, a test of the model was carried out using annual data over the period 1947 to 1984. The estimated model consisted of a system of five simultaneous equations, linear in the variables but highly nonlinear in the parameters. Hence numerical optimization was applied to obtain the reported maximum likelihood parameter estimates. Due to the LRE modeling, it was possible to distinguish between the structural parameters of the model and the expectations formation parameters which relate to the stochastic environment of the agents' problems. As a result of a number of technically and numerically difficult to be resolved problems that emerged from the estimation technique, a constrained version of the model was eventually estimated. Despite the complexities arising from the nonlinear optimization process and the model's structure requiring the computation of eigenvalues and eigenvectors at every iteration step, a relatively good fit of the model was reached. All parameter estimates were highly significant and in agreement with the theoretical predictions of the model. The main findings of this empirical test are summarized below.

One of the parameters of interest was the coefficient of the cumulative production of crude in the production cost function of the representative domestic producer of crude petroleum. The sign of this parameter could be interpreted in two ways: a positive value could imply the exhaustible nature of crude petroleum while a negative one the cost saving effect of experience and technological progress. The estimate of this coefficient turned out to be positive providing support for the Epple and Hansen (1981) exhaustible resource supply model. Among the other parameters, a steeply increasing marginal cost of holding inventories of crude petroleum and refined products by the representative domestic refiner was estimated. Furthermore, the estimated rate of increase of the marginal production costs of the same producer was even higher. These values of the curvature parameters of the cost functions imply an incentive for the refiners to smooth production in the case of fluctuating demand. It should be noted that estimates of such parameters could not be found in previous studies and thus it was not possible to provide any counterevidence.

The estimates of the characteristic roots associated with the law of motion of the inventories of crude petroleum and refined products held by the domestic refiner and the law of motion of the cumulative production of crude petroleum implied a monotonic adjustment of the inventories to changes in the exogenous variables over time. In particular, fast rates of adjustment of the inventories of crude petroleum and refined petroleum products were predicted over a year. This evidence is in accordance with the stylized facts pertaining to the observed data and also with existing evidence, although no clear conclusion can be drawn on this point given the different data samples and procedures used by earlier studies as detailed in Section 3. However, according to the predictions of this study, it is safe to argue that the inventory behavior in the refining sector does not seem to be against the production smoothing model. On the other hand and also conforming to prior belief, a slow adjustment speed was estimated for the cumulative production of crude.

Another important result of the paper is the upward sloping foreign supplies of crude and refined petroleum that implies some market power of the United States in the international petroleum market. With respect to the refined petroleum products market, a relatively high elasticity of the domestic demand for the composite of refined products was predicted. The estimate of the short run elasticity of this demand was relatively close but nevertheless larger than reported estimates of earlier studies. This gives credence to the theory that high prices of crude petroleum induce strong conservation measures in the U.S.

Overall, the diagnostic statistics for each individual equation showed a relatively good fit except for the problem of serial correlation that seems to be present in the refined products price equation and the crude petroleum inventory equation. It should also be mentioned that a test of the cross-equation restrictions implied by the theoretical model was rejected in favor of an unrestricted distributed-lag version of the model. This result was not unexpected, however, given the highly complicated set of restrictions. On the other hand, there is no clear evidence as to what is the appropriate statistic for the test of this hypothesis.

In concluding this study, it is perhaps appropriate to discuss some of the caveats of the model and make suggestions for possible modifications in the future. These modifications can apply either on the theoretical part of the model, or on its empirical implementation. In terms of the theoretical model, there are a number of factors left out of the formulation that could work in the direction of further improving the flexibility of the model for policy purposes.

Along the same lines, the behavior of the domestic government was taken to be nonstrategic. An interesting extension of the model would be to model the optimizing behavior of the government within a game theoretic framework. This could be done, for example, along the lines of Hansen, Epple and Roberts (1985). Other possible extensions can apply in the production technologies of the domestic producers of crude petroleum and refined products. In particular, as already pointed out in Section 2, some useful extensions could be the reformulation of the cost functions (5)-(6) to account for (a) other inputs explicitly such as labor and capital resources, (b) internal adjustment costs such as those related to taking delivery and preparing for the productive implementation of new orders, and (c) inventory adjustment costs as well as potential interrelations between adjustment and inventory holding costs, and (d) backlog costs. Some of these extensions have been undertaken by Blanchard (1983), Eichenbaum (1983, 1984), Kollintzas and Husted (1984), Maccini (1984) and West (1986). In addition contracts in both ends of the market could be modelled in a way similar to the "time to build" models.

On the econometric front it should be interesting to test the model with quarterly data, which we were unable to obtain. This may give greater flexibility in correcting the autocorrelation problem. On the other hand, there are several potential uses for the existing model. First, of course, the model is suitable for prediction and policy evaluation purposes. Second, the model may be used to compare alternative equilibria (i.e., market structures) in terms of their goodness of fit. Third, the Sargent-Sims debate notwithstanding, the model can be used to design optimal government policy.[19]

---

[19] Sargent (1984).

Appendix A: Mathematical Appendix

A.1 Proof of Proposition 1: (The "if" part)

Suppose that:

$$V_r^*[\bar{C}, \bar{R}; \tau] = \max_{\{C_{dd}(t), R_{dp}(t), R_{ds}(t)\}_{t=\tau}^{\infty} \in S_r[\bar{C}; \bar{R}; \tau]} V_r[\{C_{dd}(t), R_{dp}(t), R_{ds}(t)\}_{t=\tau}^{\infty}, \bar{C}, \bar{R}; \tau] \quad (A.1)$$

and

$$V_c^*[\bar{S}, \tau] = \max_{\{C_{ds}(t)\}_{t=\tau}^{\infty} \in S_c[\bar{S}, \tau]} V_c[\{C_{ds}(t)\}_{t=\tau}^{\infty}, \bar{S}, \tau] \quad (A.2)$$

where:

$$S_r[\bar{C}, \bar{R}; \tau] = \{\{C_{dd}(t), R_{dp}(t), R_{ds}(t)\}_{t=\tau}^{\infty} \mid C(t+1) = C(t) + C_{dd}(t) - \alpha R_{dp}(t);$$

$$C(\tau) = \bar{C} > 0; \quad R(t+1) = R(t) + R_{dp}(t) - R_{ds}(t); \quad R(\tau) = \bar{R} > 0;$$

$$C_{dd}(t), R_{dp}(t), R_{ds}(t), C(t+1), R(t+1) \geq 0, \forall t \in \{\tau, \tau+1, \ldots\}\}$$

$$S_c[\bar{S}; \tau] = \{\{C_{ds}(t)\}_{t=\tau}^{\infty} \mid S(t+1) = S(t) + C_{ds}(t); \quad S(\tau) = \bar{S} > 0;$$

$$C_{ds}(t), S(t+1) \geq 0, \forall t \in \{\tau, \tau+1, \ldots\}\}$$

exist and are finite. Further, suppose that the supremums in (A.1) and (A.2) are attained for some $\{C_{dd}^*(t), R_{dp}^*(t), R_{ds}^*(t)\}_{t=\tau}^{\infty} \in \overset{o}{S}_r(C, R; \tau)$ and $\{C_{ds}^*(t)\}_{t=\tau}^{\infty} \in \overset{o}{S}_c(C, R, \tau)$, where $\overset{o}{S}_r$ and $\overset{o}{S}_c$ denote the interiors of S_r and S_c, respectively. That is, $C_{dd}^*(t), R_{dp}^*(t), R_{ds}^*(t), R^*(t+1), C^*(t+1),$

$C^*_{ds}(t)$, $S^*(t+1) > 0$, $\forall t \in \{\tau, \tau+1, \ldots\}$. then, in view of the properties of $\{\Omega_t\}_{t=\tau}^{\infty}$ it follows by the Principle of Optimality that: $(C^*_{dd}(t), R^*_{dp}(t), R^*_{ds}(t))$ is a solution to

$$V^*_r[C^*(t), R^*(t); \tau] = \max_{C_{dd}(t), R_{dp}(t), R_{ds}(t) > 0} \{[p_r(t) - q_r(t)]$$

$$R_{ds}(t) - p_c(t)C_{dd}(t) - \Gamma[R_{dp}(t); t] - \Delta[C^*(t), R^*(t); t]$$

$$+ \beta E V^*_r [C^*(t) + C_{dd}(t) - \alpha R_{dp}(t), R^*(t) + R_{dp}(t) - R_{ds}(t); t]\} \quad (A.3)$$

and $C^*_{ds}(t)$ is a solution to:

$$V^*_c[S^*(t); t] = \max_{C_{ds}(t) > 0} \{[p_c(t) - q_c(t)]C_{ds}(t) - E\{C_{ds}(t), S^*(t); t]$$

$$+ \beta E V^*_c[C_{ds}(t) + S^*(t); t]\} \quad (A.4)$$

for all $t \in \{\tau, \tau+1, \ldots\}$.

Now, since $[p_r(t) - q_r(t)]R_{ds}(t) - p_c(t)C_{dd}(t) - \Gamma[R_{dp}(t); t] - \Delta[C(t), R(t); t]$ and $[p_c(t) - q_c(t)]C_{ds}(t) - E[C_{ds}(t), S(t), t]$ are differentiable and concave in $(C_{dd}(t), R_{dp}(t), R_{ds}(t), C(t), R(t))$ and $(C_{ds}(t), S(t))$, respectively; and since $\{C^*_{dd}(t), R^*_{dp}(t), R^*_{ds}(t)\}_{t=\tau}^{\infty} \in \mathcal{S}_r(\overline{C}, \overline{R}; \tau)$ and $\{C^*_{ds}(t)\}_{t=\tau}^{\infty} \in \mathcal{S}_c$, it follows that $V^*_r[C^*(t), R^*(t); t]$ and $V^*_c[S^*(t); t]$ are differentiable. See, Theorem 7.7 in Stokey, Lucas, and Prescott (1987).

Thus, we must have:

$$p_c(t) = \beta E_t\left[\frac{\partial V^*_r}{\partial C(t+1)}\bigg|^{t+1}\right] \quad (A.5)$$

$$-\frac{\partial \Gamma}{\partial R_{dp}(t)}\bigg|^t = \beta E_t\left[\frac{\partial V^*_r}{\partial C(t+1)}\bigg|^{t+1} \alpha - \frac{\partial V^*_r}{\partial R(t+1)}\bigg|^{t+1}\right] \quad (A.6)$$

$$p_r(t) - q_r(t) = \beta E_t\left[\frac{\partial V_r^*|^{t+1}}{\partial R(t+1)}\right] \qquad (A.7)$$

$$p_c(t) - q_c(t) - \frac{\partial \Theta|^t}{\partial C_{ds}(t)} = -\beta E_t\left[\frac{\partial V_c^*|^{t+1}}{\partial S(t+1)}\right] \qquad (A.8)$$

where: $\dfrac{\partial V_r^*|^{t+1}}{\partial C(t+1)}$ stands for the partial derivative of $V_r^*[C(t+1), R(t+1), t+1]$ evaluated at $(C^*(t+1), R^*(t+1), t+1)$, etc. Further, we must have:

$$\frac{\partial V_r^*|^{t+1}}{\partial C(t+1)} = -\frac{\partial \Delta|^{t+1}}{\partial C(t+1)} + \beta E_{t+1}\left[\frac{\partial V_r^*|^{t+2}}{\partial C(t+2)}\right] \qquad (A.9)$$

$$\frac{\partial V^*|^{t+1}}{\partial R(t+1)} = -\frac{\partial \Delta|^{t+1}}{\partial R(t+1)} + \beta E_{t+1}\left[\frac{\partial V_r^*|^{t+2}}{\partial R(t+2)}\right] \qquad (A.10)$$

$$\frac{\partial V_c^*|^{t+1}}{\partial S(t+1)} = -\frac{\partial \Theta|^{t+1}}{\partial S(t+1)} + \beta E_{t+1}\left[\frac{\partial V_c^*|^{t+2}}{\partial S(t+2)}\right] \qquad (A.11)$$

In view of (A.5)-(A.8), (A.9)-(A.11) give:

$$p_c(t) = \beta E_t\left[-\frac{\partial \Delta|^{t+1}}{\partial C(t+1)} + p_c(t+1)\right] \qquad (A.12)$$

$$p_r(t) - q_r(t) = \beta E_t\left[-\frac{\partial \Delta|^{t+1}}{\partial R(t+1)} + p_r(t+1) - q_r(t+1)\right] \qquad (A.13)$$

$$p_c(t) - q_c(t) - \frac{\partial \Theta|^t}{\partial C_{ds}(t)} = \beta E_t\left[\frac{\partial \Theta|^{t+1}}{\partial S(t+1)}\right]$$

$$+ p_c(t+1) - q_c(t+1) - \frac{\partial \Theta|^{t+1}}{\partial C_{ds}(t+1)} \tag{A.14}$$

or dropping the *-notation,

$$p_c(t) = \beta E_t[-\beta^{-1} d_1(t) + d_1(t+1) - d_2 C(t+1)$$

$$- d_3 R(t+1) + p_c(t+1)] \tag{A.15}$$

$$p_r(t) - q_r(t) = \beta E_t[-d_3 C(t+1) - d_4 R(t+1)$$

$$+ p_r(t+1) - q_r(t+1)] \tag{A.16}$$

$$p_c(t) - q_c(t) - [e_1(t) + e_2 C_{ds}(t) + e_3 S(t)] = -\beta E_t[-e_3 C_{ds}(t+1)$$

$$- p_c(t+1) + q_c(t+1) + e_1(t+1) + e_2 C_{ds}(t+1) + e_3 S(t+1)] \tag{A.17}$$

In addition, if $\{p_c(t), p_r(t), n_c C_{ds}(t), n_r R_{dp}(t), n_r R_{ds}(t), v(t)\}_{t=\tau}^{\infty}$ is a REE we must have:

$$n_r R_{ds}(t) + R_{fs}(t) = \frac{1}{a_2} [a_1(t) - p_r(t) + a_3 p_s(t) + a_4 Y(t)] \tag{A.18}$$

$$R_{fs}(t) = b_1(t) + b_2[p_r(t) - r_r(t)] \tag{A.19}$$

$$n_c C_{ds}(t) + C_{fs}(t) = n_r C_{dd}(t) + X(t) \tag{A.20}$$

$$C_{fs}(t) = \begin{cases} \dfrac{1}{f_2}[f_1(t) + p_c(t) - r_c(t)], & \text{without quota} \\ \\ w(t) & \text{, with quota} \end{cases} \tag{A.21}$$

Moreover, (A.5) and (A.7) imply that:

$$-c_1(t) - c_2 R_{dp}(t) = \alpha\, p_c(t) - p_r(t) + q_r(t) \qquad (A.22)$$

(A.18)-(A.22) imply:

$$p(t) = Q\, u(t) + P\, v(t) \qquad (A.23)$$

where:

$$Q = \begin{bmatrix} 0 & -c_2/\alpha & -a_2/\alpha g \\ 0 & 0 & -a_2/g \end{bmatrix}$$

$$P = \begin{bmatrix} 0 & -1/\alpha & 0 & a_2 b_2/\alpha g & a_3/\alpha g & 0 & 0 & a_4/\alpha g & 1/\alpha g \\ 0 & 0 & 0 & a_2 b_2/g & a_3/g & 0 & 0 & a_4/g & 1/g \end{bmatrix}$$

$$\begin{bmatrix} -a_2/\alpha g & -1/\alpha & 0 & 0 & 0 \\ -a_2/g & 0 & 0 & 0 & 0 \end{bmatrix}, \quad g = (1 + a_2 b_2)$$

Furthermore, in view of (A.18)-(A.22), (A.15)-(A.17) give:

$$\beta^{-1} D\, u(t) + \beta^{-1} B\, v(t) = -F\, E_t\, x(t+1) + DE_t u(t+1) + B_t E\, v(t+1) \qquad (A.24)$$

where:

$$D = \begin{bmatrix} 0 & -n_r^{-1} c_2/\alpha & -a_2/\alpha g \\ 0 & 0 & -a_2/g \\ n_c^{-1}(e_2 - e_3) & n_r^{-1} c_2/\alpha & a_2/\alpha g \end{bmatrix}$$

$$B = \begin{bmatrix} 0 & -1/\alpha & 0 & a_2 b_2/\alpha g & a_3/\alpha g & 0 & 0 & a_4/\alpha g \\ 0 & -1 & 0 & a_2 b_2/g & a_3/g & 0 & 0 & a_4/g \\ 1 & 1/\alpha & 0 & -a_2 b_2/\alpha g & -a_3/\alpha g & 0 & 0 & -a_4/\alpha g \end{bmatrix}$$

$$\begin{pmatrix} 1/\alpha g & -a_2/\alpha g & -1/\alpha & 1 & 0 & 0 \\ 1/g & -a_2/\alpha g & 0 & 0 & 0 & 0 \\ -1/\alpha g & a_2/\alpha g & 1/\alpha & 0 & 1 & 0 \end{pmatrix}$$

$$F = \begin{bmatrix} n_r^{-1}d_2 & n_r^{-1}d_3 & 0 \\ n_r^{-1}d_3 & n_r^{-1}d_4 & 0 \\ 0 & 0 & n_c^{-1}(\beta^{-1}-1)e_3 \end{bmatrix}$$

Finally, from the transition equations and (A.20)-(A.21), we have:

$$x(t+1) = x(t) + G\,u(t) + H\,v(t) \tag{A.25}$$

where:

$$(G,H) = \begin{cases} (G_0, H_0), & \text{without the quota} \\ (G_1, H_1), & \text{with the quota} \end{cases}$$

$$G_0 = \begin{bmatrix} 1 & -\alpha - (n_r^{-1}c_2/\alpha f_2) & -a_2/\alpha f_2 g \\ 0 & 1 & -1 \\ 1 & 0 & 0 \end{bmatrix}$$

$$G_1 = \begin{bmatrix} 1 & -\alpha & 0 \\ 0 & 1 & -1 \\ 1 & 0 & 0 \end{bmatrix}$$

$$H_0 = \begin{bmatrix} 0 & -1/\alpha f_2 & -1/f_2 & \dfrac{a_2 b_2}{\alpha f_2 g} & a_3/\alpha f_2 g & 0 & -1 & \dfrac{a_4}{\alpha f_2 g} \\ 0 & 0 & 0 & 0 & 0 & 0 & 0 & 0 \\ 0 & 0 & 0 & 0 & 0 & 0 & 0 & 0 \end{bmatrix}$$

$$\left. \begin{array}{cccccc} \dfrac{1}{\alpha f_2 g} & \dfrac{-a_2}{\alpha f_2 g} & \dfrac{-1}{\alpha f_2} & 0 & 0 & 1/f_2 \\ 0 & 0 & 0 & 0 & 0 & 0 \\ 0 & 0 & 0 & 0 & 0 & 0 \end{array} \right)$$

$$H_1 = \begin{pmatrix} 0 & 0 & 0 & 0 & 0 & 1 & -1 & 0 & 0 & 0 & 0 & 0 & 0 & 0 \\ 0 & 0 & 0 & 0 & 0 & 0 & 0 & 0 & 0 & 0 & 0 & 0 & 0 & 0 \\ 0 & 0 & 0 & 0 & 0 & 0 & 0 & 0 & 0 & 0 & 0 & 0 & 0 & 0 \end{pmatrix}$$

As shown in A.2, G is invertible. Therefore, (A.24) and (A.25) give:

$$J\, E_t\, x(t+2) - [(1+\beta^{-1})J + F]E_t\, x(t+1) + \beta^{-1} J\, E_t\, x(t)$$
$$= (B-JH)E_t\,[\beta^{-1}v(t) - v(t+1)] \qquad (A.26)$$

where

$$J = DG^{-1}$$

This proves that (19)-(23) are necessary for $\{p(t),u(t),v(t)\}_{t=\tau}^{\infty}$ to be a corner free rational expectations equilibrium. The sufficiency follows by a standard argument.

A.2 Proof of the Properties of J

Property 1: J is well defined.

Let:

$$d_{12} = \frac{-n_r^{-1}c_2}{\alpha},\quad d_{13} = \frac{-a_2}{a(1+a_2 b_2)},\quad d_{23} = \frac{-a_2}{1+a_2 b_2}$$

$$d_{31} = n_c^{-1}(e_2-e_3),\quad g_{12} = -\alpha - \frac{\phi n_r^{-1}c_2}{\alpha f_2},\quad g_{13} = \frac{-a_2}{\alpha(1+a_2 b_2)f_2}$$

where:

$$\phi = \begin{cases} 0, & G = G_0 \\ 1, & G = G_1 \end{cases}.$$

Then, it follows by definition that:

$$D = \begin{pmatrix} 0 & d_{12} & d_{13} \\ 0 & 0 & d_{23} \\ d_{31} & -d_{12} & -d_{13} \end{pmatrix}, \quad G = \begin{pmatrix} 1 & g_{12} & g_{13} \\ 0 & 1 & -1 \\ 1 & 0 & 0 \end{pmatrix}$$

Therefore, $|G| = -(g_{12} + g_{13}) < 0$, $\forall \phi \varepsilon \{0,1\}$; and $J = DG^{-1}$ is well defined.

Property 2: J is symmetric.

By definition:

$$J = \begin{pmatrix} 0 & d_{12} & d_{13} \\ 0 & 0 & d_{23} \\ d_{31} & -d_{12} & -d_{13} \end{pmatrix} \begin{pmatrix} 1 & g_{12} & g_{13} \\ 0 & 1 & -1 \\ 1 & 0 & 0 \end{pmatrix}^{-1}$$

$$= \begin{pmatrix} 0 & d_{12} & d_{13} \\ 0 & 0 & d_{23} \\ d_{31} & -d_{12} & -d_{13} \end{pmatrix} \begin{pmatrix} 0 & 0 & g_{12}+g_{13} \\ 1 & g_{13} & -1 \\ 1 & -g_{12} & -1 \end{pmatrix} / (g_{12}+g_{13})$$

$$= \begin{pmatrix} d_{12}+d_{13} & d_{12}g_{13}-d_{13}g_{12} & -(d_{12}+d_{13}) \\ d_{23} & -d_{23}g_{12} & -d_{23} \\ -(d_{12}+d_{13}) & -(d_{12}g_{13}-d_{13}g_{12}) & (d_{12}+d_{13})+d_{31}(g_{12}+g_{13}) \end{pmatrix} / (g_{12}+g_{13})$$

$$= \begin{pmatrix} J_{11} & J_{12} & -J_{11} \\ J_{12} & J_{22} & -J_{12} \\ -J_{11} & -J_{12} & J_{33} \end{pmatrix} = J',$$

since $\quad d_{12}g_{13} - d_{13}g_{12} = -\dfrac{-a_2}{1+a_2 b_2} = d_{23}$.

Property 3: J is positive semi-definite.

Clearly, $J_{11}, J_{22}, J_{33} \geq 0$. $J_{11}J_{22} - J_{12}^2 \geq 0 \Leftrightarrow (d_{12}+d_{13})(-d_{23}g_{12}) - d_{23}^2 \geq 0 \Leftrightarrow (d_{12}+d_{13})g_{12}+d_{23} \geq 0$. But

$$(d_{12}+d_{13})g_{12}+d_{23} = \left(\dfrac{-n_c^{-1}c_2}{\alpha} - \dfrac{a_2}{\alpha(1+a_2 b_2)}\right)(-\alpha - \dfrac{\phi n_c^{-1}c_2}{\alpha f_2})$$

$$-\dfrac{a_2}{1+a_2 b_2} \geq n_c^{-1}c_2 \geq 0.$$

$J_{22}J_{33} - J_{12}^2 \geq 0$ for $J_{33} = J_{11} + d_{31} \geq J_{11}$ and we have already shown that $J_{11}J_{22} - J_{12}^2 \geq 0$. $|J| = J_{11}(J_{22}J_{33} - J_{12}^2) - J_{12}(J_{12}J_{33} - J_{11}J_{12}) - J_{11}(-J_{12}^2 + J_{11}J_{22}) = J_{11}[(J_{22}J_{33}-J_{12}^2) - (J_{11}J_{22}-J_{12}^2)] - J_{12}^2(J_{33}-J_{11}) = J_{11}J_{22}d_{31} - J_{12}^2 d_{31} = (J_{11}J_{22}-J_{12}^2)d_{31} \geq 0$. This, of course, proves that J is positive semi-definite for all its principal minors are non-negative.

Property 4: J nonsingular if and only if $a_2, c_2, e_2 - e_3 > 0$

We have already shown that:

$$|J| = (J_{11}J_{22} - J_{12}^2)d_{31}$$

Since $(J_{11}J_{22} - J_{12}^2)$, $d_{31} \geq 0$, $|J| > 0$ if and only if $J_{11}J_{22} - J_{12}^2$, $d_{31} > 0$. Clearly, $d_{31} > 0 \to e_2 - e_3 > 0$. Also, it follows as in the previous proof that $J_{11}J_{22} - J_{12}^2 > 0$. But, $J_{11}J_{22} - J_{12}^2 = -(d_{12}+d_{13})d_{23}g_{12} - d_{23}^2 = -d_{23}[(d_{12}+d_{13})g_{12}+d_{23}]$. Likewise, $J_{11}J_{22} - J_{12}^2 > 0$ if and only if $-d_{23} > 0$; $(d_{12}+d_{13})g_{12}+d_{23} > 0 - d_{23} > 0 \to a_2 > 0$. On the other hand, $(d_{12}+d_{13})g_{12} + d_{23} = n_c^{-1}c_2 + (n_c^{-1}c_2 + a_2/1+a_2 b_2) \phi n_c^{-1}c_2/\alpha^2 f_2 > 0$, $\forall \phi \varepsilon [0,1]$, if and only if $c_2 > 0$. It follows that $|J| \neq 0$ if and only if $a_2, c_2, e_2 - e_3 > 0$.

APPENDIX B
DATA ESTIMATES AND SOURCES OF DATA

A. GLOSSARY OF SYMBOLS AND SOURCES OF DATA

CINV: Beginning of period crude petroleum inventories, $C(t)$, in millions of barrels.[20] Data estimates were obtained from various issues of the *Business Statistics*, and the *Survey of Current Business*. This series was constructed by adding the series "Total crude petroleum, stocks, end of period" to the series "Unfinished oils, natural gasoline, etc, stocks, end of period". The resulting series was shifted forward by a year to convert the data to beginning of period stocks. This inventory series does not include the Strategic petroleum reserves data and is presented in the first column of Table 1.

COUT: Domestic production of crude petroleum, $C_{ds}(t)$, in millions of barrels. Data estimates for the period 1860-1947 were obtained from the U.S. Department of Commerce, Historical Statistics of the United States, Bicentennial Edition. For the years 1947 through 1984, these data were taken from various issues of the *Business Statistics* and the *Survey of Current Business*, and were constructed by adding the series "Crude petroleum, production, all oils, new supply" to the series "Natural gas plant liquids, production, all oils, new supply." The data for the period 1947 to 1984 are presented in the second column of Table 1.

CPCU: Beginning of period cumulative U.S. production of crude petroleum $S(t)$, in millions of barrels. This series was constructed using the relation [see equation (10)]:

$$S(t+1) = S(t) + C_{ds}(t) \; ; \; S(1860) = 0$$

This series is presented in the last column of Table 1.

RINV: Beginning of period inventories of refined petroleum products, $R(t)$, in millions of barrels. These data were obtained from

---

[20] Throughout this appendix a barrel is referred to as U.S. barrel of 42 gallons.

various issues of the Business Statistics and the Survey of Current Business. The series "Refined products, stocks, end of period" was shifted forward by a year to convert the data to beginning of period stocks and is presented in the first column of Table 2.

ROUT: Domestic production of refined petroleum products, $R_{dp}(t)$, in millions of barrels. This series was constructed by adding the series: "Production, gasoline (including aviation)", "Production, kerosene", "Production, distillate fuel oil", "Production, residual fuel oil", "Production, jet fuel", "Production, lubricants", "Production, asphalt", and "Total production, liquified gases (including ethane and methane). All series were taken from various issues of the Business Statistics and the Survey of Current Business. The resulting series is presented in the second column of Table 2.

CIMP: Net imports of crude petroleum, $C_{fs}(t)$, in millions of barrels. Estimates of this series were obtained from various issues of the Business Statistics and the Survey of Current Business by subtracting the series "Crude petroleum, exports, demand, all oils, demand and stocks" from the series "Crude petroleum and unfinished oils, imports, all oils, new supply". The resulting series is presented in Table 3.

RIMP: Net imports of refined petroleum products, $R_{fs}(t)$, in millions of barrels. The data sources for this series are also various issues of the Business Statistics and the Survey of Current Business. These data were constructed by subtracting the series "Refined products, exports, demand, all oils, demand and stocks" from the series "Refined products, imports, all oils, new supply". These data are presented in Table 3.

CONR: Domestic consumption of crude petroleum, $C_{dc}(t)$, in millions of barrels. The data estimates of this series were obtained from the relation [see equations (1) and (2)]:

$$C_{dc}(t) = C_{dd}(t) - [C(t) - C(t-1)]$$

$$= C_{dp}(t) + C_{fs}(t) - [C(t) - C(t-1)]$$

The resulting series is presented in Table 1.

RSAL: Sales of refined petroleum products, R_{ds}, in millions of barrels. This series was constructed according to the relation [see equation (2)]:

$$R_{ds}(t) = R_{dp}(t) - [R(t) - R(t-1)]$$

The resulting series is presented in Table 2.

CDPRI: Domestic producers' price of crude petroleum, $p_c(t) \cdot q_c(t)$, in dollars per barrel. The sources used to obtain these prices are the Bureau of Labor Statistics, <u>Wholesale Prices and Price Indexes</u>. The series was constructed by multiplying the wholesale price index of crude petroleum (Table 4) by the dollar value of crude of the base year (1967=100). The latter was computed by weighting the prices of the different types of crude by the relative importance of each type included in the wholesale price index. The price index of crude (1967=100) is also reported in the <u>Business Statistics</u> and the <u>Survey of Current Business</u>, and is based on buyers posted prices (obtained from the petroleum companies) of crude petroleum produced in thirteen areas in U.S. Crude petroleum prices are presented in Table 7.

RDPRI: Domestic producers' price of refined petroleum products, $p_r(t) \cdot q_r(t)$, in dollars per barrel. This series is a weighted average of the prices of the four major refined petroleum products: gasoline, kerosene (light distillate fuel oil), distillate (middle distillate) fuel oil, and residual fuel oil for which price data are readily available. As indicated in Table 6, these four products account for over 77 percent of the yields of refined products from a barrel of crude. This fact justifies the reason we are using the prices of these four products only, given that prices on the other petroleum products are not readily available. The prices of the above-mentioned four products were computed in the same way the price of crude was computed. The Wholesale (Producer) Price Index of each type of product and the corresponding dollar value of the base year were also taken from the Bureau of Labor Statistics, <u>Wholesale Prices and Price Indexes</u>. The producer price index (1967=100) of kerosene, distillate fuel oil and residual fuel oil are also reported in the <u>Business Statistics</u> and the <u>Survey of Current Business</u> as well, but for gasoline (regular grade) only the updated index (Feb. 1973=100) is mentioned there. As weights for the estimate of the composite refined products, the domestic sales given by the series "Domestic product, demand, all oils, demand and stocks" of the Business Statistics and the Survey of Current Business were used (Table 5). The indexes of the refined petroleum products are presented in Table 4 and the prices in Table 7.

CGIM: U.S. general imports of crude petroleum in millions of barrels. Data estimates for the period 1947-1963 were obtained from the annual issues of U.S. Department of Commerce, FORM 110, United States Imports of Merchandize for Consumption. This series was constructed by adding the total net quantity, Schedule A, of crude petroleum (testing under 25 degrees API and 25 degrees API or more) to the total of topped crude and unfinished oils. For the years following 1963, the data were taken from the December issues of the U.S. Department of Commerce, FORM 125, United States Imports of Merchandize for Consumption (1964-1966) and FORM 135, U.S. General Imports, Schedule A, Commodity by Country of Origin (1967-1984). Starting February 1963, the data in these publications were presented in terms of the Standard International Trade Classification (SITC) [see notes under Table 9]. Under the new classification the series for the period 1964 through 1984 was constructed by summing the total of crude petroleum testing under 25 degrees API, "net quantity, Cumulative, January to date, Schedule A" to the total of crude petroleum testing 25 degrees API or over, "net quantity, Cumulative, January to date, Schedule A". These data are reported in Table 10. The resulting series is very similar to the series of crude petroleum imports provided by the Survey of Current Business (Table 3). The reason of constructing this new series of imports is to obtain the corresponding value of U.S. imports which is not reported by the Survey of Current Business.

RGIM: U.S. general imports of refined petroleum products in millions of barrels. This series was constructed using the same sources as those used before to obtain the relevant series for crude petroleum. All the reported subgroups of refined petroleum products were considered, including the imports of liquified petroleum gases for consistency with the classification of the Survey of Current Business.

For those products whose imports were not measured in barrels, Table 9 was used to convert all units of measurement to U.S. barrels (42 gallons) according to the international standards. These import data on refined petroleum products are reported in Table 11.

CFPRI: Price of imported crude petroleum, $p_c(t) - r_c(t)$, in dollars per barrel. Data estimates of these prices were constructed by dividing the dollar value of imports of crude petroleum (millions of dollars) by the corresponding volume of general imports (millions of barels) previously defined as CGIM (Table 10). For reasons of data consistency both series were taken from the same sources as in CGIM before. The corresponding import value reported in these sources represents either customs or f.a.s. (free alongside ship) value which excludes U.S. import duties, freight, insurance and other charges incurred in bringing the merchandize in the U.S. Beginning January 1974, these statistics report in addition the c.i.f. (cost, insurance, freight) value which represents the value of imports at the first port of entry in the U.S., but still excludes U.S. import duties. The value used to compute our series of imported price was the c.i.f. value for 1974 onward, while for the previous years the f.a.s. value was used to extrapolate the c.i.f. series backwards to 1947. The original import value series along with the extrapolated one and also the derived value of imported crude petroleum are reported in Table 9.

RFPRI: Price of imported refined petroleum products, $p_r(t) - r_r(t)$, in dollars per barrel. Data estimates of these prices were constructed as before by dividing the equivalent c.i.f. dollar value of general imports by the series RGIM defined previously. The c.i.f. values prior to 1974 are also derived from the corresponding f.a.s. values by extrapolation. These price estimates are shown in Table 11.

It can be noticed from Table 12 that the estimated price series of refined products RFPR does not differ substantially from the estimated crude price series CFPR, although the former was expected to be bigger by the refining costs. This is not surprising, however, since the bulk of imported products consists of residual fuel which is relatively cheap as compared to gasoline which is the most expensive product but its role in the import market is very small. Moreover, a large part of US refined product imports has historically come from refining operations in near-neighbors such as Canada, Mexico, and the Central America and Caribbean area, which may partially account for the low refined product prices. Yet, no definite conclusion can be derived given that the crude imports from these neighbors have also increased recently in an attempt to reduce dependence on the Arab world.

As a more representative estimate of the refined products price the series in Table 13 was used. This was constructed as a weighted average of the retail price of gasoline including state and federal excise taxes reported in the same table and the wholesale prices of the other three products (Table 4) since no readily available data on the retail prices of these three products could be found for the period before 1978. The retail price of gasoline for 1947-1977 was taken from the 1976 issue of the American Petroleum Institute: <u>Basic Petroleum Data Book, Petroleum Statistics</u>, while for the remaining years the summary statistics from the <u>Petroleum Marketing Monthly</u> were used. As weights the estimates of the domestic demand for each of these products were used (Table 5).

QC: The tax (or subsidy) imposed on crude petroleum, $q_c(t)$, in dollars per barrel. It includes the federal corporation income tax, any existing excise taxes, and an estimate of the federal government regulation that kept domestic prices different from their world levels, as discussed in section 2. For an estimate of the first one, an **ad valorem** federal corporate income tax rate was constructed by dividing the series "Total Income Tax" by the series "Business Receipts" in the oil and gas extraction industry. Both series were taken from various issues of the U.S. Treasury Department, Internal Revenue Service, "Statistics of Income - Corporation Income Tax". This tax rate was then applied to the wholesale domestic price of crude (Table 7) to compute the relevant tax. No excise taxes on crude oil existed until 1980 when the Windfall Profits Tax became effective in order to phase out the existing controls. Table 14 gives these tax collections as reported in the <u>Statistical Abstract of the U.S.</u>, 1986, and also the estimated tax rate as a percent of the "Gross Receipts" of this industry. Finally, as an estimate of the federal regulation, the differences between the world and the domestic prices of crude were used. Since this last estimate theoretically includes all excise taxes, the Windfall Profits Tax was not eventually included to avoid double counting. The tax rates and the final estimate of QC in real terms are reported in Table 15. It should be mentioned that state taxes were not considered since no collective data were readily available and the task of computing such rates from raw data at the state level was beyond the scope of this study.

QR: The tax (or subsidy) imposed on refined petroleum products, $q_r(t)$, in dollars per barrel. As before, QR includes an estimate of the federal corporation income tax and an estimate of the federal government regulation (excise and other taxes). To compute the former, an estimate of the **ad valorem** tax rate was constructed by dividing the "Total Income Tax" by the "Business Receipts" of the oil and gas extraction industry (see sources in the previous paragraph). This rate was then applied to the wholesale prices of refined products (Table 7) to estimate the relevant tax income. The latter was also computed as the difference between the estimated world and domestic price of the composite refined products series. No additional excise taxes were computed since the estimated regulation implicitly includes all types of excise taxes. Alternatively, QR was estimated by taking the federal corporation income tax as before and the existing federal and state excise taxes. This series was used in combination with the new series of refined product prices presented in Table 13. The federal and state taxes for gasoline were taken from the 1976 issue of the American Petroleum Institute: <u>Basic Petroleum Data Book: Petroleum Statistics</u>. It was estimated that these taxes have behaved like a sales tax at the rate of 27.3175 percent which was used to estimate the taxes for the years 1978 through 1984. This series is reported in Table 15.

The input/output ratio α: It was obtained using the relation

$$C_{dc}(t) = \alpha R_{dp}(t)$$

See footnote 15 for the OLS estimate of α.

Table B-1: U.S. Crude Petroleum Stocks, Production and Cumulative Production Estimates, 1947 to 1984

(Millions of barrels)

Year	CINV Inventories $C(t)$	COUT Domestic Production $C_{ds}(t)$	CRCU Cumulative Production $S(t)$
1947	235.70	1,989.90	34,497.10
1948	235.00	2,167.30	36,487.00
1949	262.20	1,999.20	38,654.30
1950	260.20	2,155.70	40,653.50
1951	255.90	2,452.70	42,809.20
1952	264.00	2,513.70	45,261.90
1953	279.70	2,596.20	47,775.60
1954	284.80	2,567.60	50,371.80
1955	272.40	2,766.30	52,939.40
1956	279.20	2,910.50	55,705.70
1957	286.60	2,912.10	58,616.20
1958	303.40	2,744.20	61,528.30
1959	285.50	2,895.70	64,272.50
1960	282.00	2,915.80	67,168.20
1961	268.70	2,983.70	70,084.00
1962	281.80	3,049.00	73,067.70
1963	283.40	3,153.70	76,116.70
1964	271.10	3,209.30	79,270.40
1965	265.80	3,290.10	82,479.70
1966	256.20	3,496.50	85,769.80
1967	278.80	3,730.20	89,266.30
1968	345.00	3,882.70	92,996.50
1969	371.10	3,956.30	96,879.20
1970	368.70	4,129.60	100,835.50
1971	382.40	4,077.80	104,965.10
1972	366.40	4,103.70	109,042.90
1973	347.20	4,006.00	113,146.60
1974	349.50	3,831.80	117,152.60
1975	378.60	3,666.50	120,984.40
1976	385.10	3,577.20	124,650.90
1977	404.10	3,618.10	128,228.10
1978	469.40	3,769.60	131,846.20
1979	493.00	3,715.50	135,615.80
1980	562.30	3,738.20	139,331.30
1981	674.90	3,734.10	143,069.50
1982	771.10	3,741.80	146,803.60
1983	801.70	3,759.20	150,545.40
1984	884.40	3,819.10	154,304.60
1985	934.00	-	158,123.70

Table B-2: U.S. Stocks, Production, Crude Input and Sales of Refined Petroleum Products, 1947 to 1984

Year	RINV Inventories $R(t)$	ROUT Domestic Production $R_{dp}(t)$	CONR Refiners Crude Consumption $C_{dc}(t)$	RSAL Refiners Sales $R_{ds}(t)$
1947	224.40	1,865.40	2,041.70	1,878.20
1948	211.60	2,060.80	2,229.50	2,002.10
1949	270.30	1,994.20	2,121.80	1,990.30
1950	274.20	2,164.30	2,302.90	2,179.30
1951	259.20	2,452.80	2,595.10	2,412.10
1952	299.90	2,556.20	2,680.90	2,535.40
1953	320.70	2,650.10	2,807.70	2,611.80
1954	359.00	2,650.00	2,805.90	2,654.20
1955	354.80	2,861.30	3,033.30	2,855.00
1956	361.10	3,027.10	3,216.30	2,964.20
1957	424.00	3,006.50	3,218.40	2,963.20
1958	467.30	2,939.50	3,105.80	2,979.00
1959	427.80	3,088.10	3,249.00	3,059.00
1960	456.90	3,133.50	3,297.60	3,142.40
1961	448.00	3,184.70	3,348.90	3,148.60
1962	484.10	3,286.20	3,456.60	3,287.90
1963	482.40	3,344.90	3,577.00	3,330.60
1964	496.70	3,422.70	3,651.80	3,422.80
1965	496.60	3,517.70	3,750.60	3,514.50
1966	499.80	3,676.30	3,919.50	3,654.30
1967	521.80	3,931.00	4,049.10	3,873.00
1968	579.80	4,143.50	4,356.50	4,115.10
1969	608.20	4,269.90	4,510.20	4,285.20
1970	592.90	4,396.00	4,633.50	4,372.40
1971	616.50	4,552.00	4,751.90	4,511.40
1972	657.10	4,734.90	4,979.50	4,800.20
1973	591.80	5,001.00	5,237.20	4,954.10
1974	638.70	4,870.20	5,115.00	4,832.50
1975	676.40	4,944.60	5,169.10	4,891.90
1976	729.10	5,246.10	5,502.00	5,290.00
1977	685.20	5,562.40	5,960.10	5,426.90
1978	820.70	5,608.90	6,018.00	5,669.00
1979	760.60	5,524.80	5,961.40	5,524.80
1980	760.60	5,130.40	5,466.90	5,166.40
1981	724.60	4,939.60	5,208.90	4,967.80
1982	696.40	4,791.30	4,977.30	4,876.50
1983	611.20	4,732.60	4,934.40	4,789.20
1984	554.60	4,963.50	5,061.70	4,911.00
1985	607.10	-	-	-

Table B-3: U.S. Imports, Exports, and Net Imports of Crude Petroleum and Refined Petroleum Products, 1947 to 1984

(Millions of Barrels)

Year	Crude Petroleum Imports	Exports	CIMP Net Imports $C_{fs}(t)$	Refined Products Imports	Exports	RIMP Net Imports $R_{fs}(t)$
1947	97.5	46.4	51.1	61.9	118.1	-56.2
1948	129.1	39.7	89.4	59.1	94.9	-35.8
1949	153.7	33.1	120.6	81.9	86.3	-4.4
1950	177.7	34.8	142.9	132.5	76.5	56.0
1951	179.1	28.6	150.5	129.1	125.4	3.7
1952	209.6	26.7	182.9	138.9	131.5	7.4
1953	236.5	19.9	216.6	141.0	126.7	14.3
1954	239.5	13.6	225.9	144.5	116.1	28.4
1955	285.4	11.6	173.8	170.1	122.6	47.5
1956	341.8	28.6	313.2	183.8	128.8	55.0
1957	373.3	50.2	323.1	201.3	156.9	44.4
1958	348.0	4.3	343.7	272.6	96.3	176.3
1959	352.3	2.5	349.8	297.2	74.5	222.7
1960	371.6	3.1	368.5	292.5	70.8	221.7
1961	381.5	3.2	378.3	318.1	60.3	257.8
1962	411.0	1.8	409.2	348.8	59.6	289.2
1963	412.7	1.7	411.0	362.1	74.2	287.9
1964	438.6	1.4	437.2	388.1	72.5	315.6
1965	452.0	1.1	450.9	448.7	67.2	381.5
1966	447.1	1.5	445.6	492.0	70.9	421.1
1967	411.6	26.5	385.1	514.3	85.5	428.8
1968	501.7	1.8	499.9	537.7	82.7	455.0
1969	552.9	1.4	551.5	602.7	83.4	519.3
1970	522.6	5.0	517.6	725.5	89.5	636.0
1971	658.6	.5	658.1	774.3	81.3	693.0
1972	856.8	.2	856.6	878.5	81.2	797.3
1973	1,234.2	.7	1,233.5	1049.3	83.7	965.6
1974	1,313.4	1.1	1,312.3	917.6	79.4	838.2
1975	1,511.2	2.1	1,509.1	699.2	74.3	624.9
1976	1,946.7	2.9	1,943.8	729.7	78.7	651.0
1977	2,425.6	18.3	2,407.3	789.1	70.3	718.8
1978	2,329.7	57.7	2,272.0	722.9	74.3	648.6
1979	2,400.9	85.7	2,315.2	685.6	86.1	599.5
1980	1,946.2	104.9	1,841.3	582.5	94.3	488.2
1981	1,654.2	83.2	1,571.0	534.2	133.9	400.3
1982	1,352.4	86.3	1,266.1	514.0	211.2	302.8
1983	1,317.8	59.9	1,257.9	525.9	209.9	316.0
1984	1,358.4	66.2	1,292.2	610.2	196.9	413.3

Table B-4: Indices of U.S. Wholesale Prices of Crude Petroleum and Refined Petroleum Products, 1947 to 1984

(Millions of Barrels)

Year	Crude Petroleum (1967=100)	Refined Petroleum Products			
		Gasoline Regular (2/73=100)	Kerosene (1967=100)	Distillate Fuel Oil (1967=100)	Residual Fuel Oil (1967=100)
1947	62.6	64.4	64.3	65.2	94.1
1948	84.3	78.1	86.0	90.3	125.6
1949	83.2	79.3	74.0	76.8	74.5
1950	83.2	80.7	78.2	80.1	86.8
1951	83.6	84.8	82.5	86.5	97.3
1952	83.6	84.7	84.0	87.1	87.2
1953	89.7	88.9	83.9	90.0	85.7
1954	92.3	84.8	85.2	91.2	91.6
1955	92.4	84.7	87.1	93.5	102.8
1956	92.9	87.1	92.2	97.8	117.0
1957	102.4	91.3	94.7	103.2	138.8
1958	102.6	85.3	88.0	94.6	109.6
1959	99.2	84.9	91.2	96.1	102.9
1960	98.6	85.7	89.7	90.5	109.7
1961	98.9	85.6	93.6	94.9	113.3
1962	99.1	84.8	93.2	93.6	111.5
1963	98.7	83.9	92.9	93.9	107.6
1964	98.3	80.2	85.2	86.5	104.8
1965	98.2	82.6	90.4	91.9	107.7
1966	98.9	87.3	93.2	93.7	105.0
1967	100.0	88.4	100.0	100.0	100.0
1968	100.8	84.8	101.0	101.9	95.7
1969	105.2	86.5	100.0	102.4	93.3
1970	106.1	89.6	102.3	106.5	125.5
1971	113.2	89.6	105.7	110.0	166.0
1972	113.8	92.0	106.7	111.3	158.8
1973	126.0	109.9	128.0	139.7	190.4
1974	211.8	178.4	226.7	272.0	485.4
1975	245.7	211.8	285.6	309.4	495.5
1976	253.6	233.6	312.3	337.0	452.9
1977	274.2	253.6	358.5	384.1	522.5
1978	300.1	265.0	392.7	398.0	498.0
1979	376.5	367.6	539.6	573.9	684.5
1980	556.4	576.7	863.4	850.6	961.2
1981	803.5	666.0	1,039.8	1,058.1	1,239.0
1982	733.4	612.5	996.4	1,012.7	1,182.0
1983	681.4	551.7	906.1	889.8	1,058.9
1984	670.5	515.5	870.0	880.2	1,120.1

Table B-5: Domestic Demand for Gasoline, Kerosene, Distillate Fuel Oil and Residual Fuel Oil in the United States, 1947 to 1984

Year	Gasoline	Kerosene	Distillate Fuel oil	Residual Fuel oil
1947	795.0	102.5	298.3	518.5
1948	871.3	112.2	340.6	500.5
1949	913.7	102.7	329.3	496.0
1950	994.3	117.8	394.9	553.8
1951	1,089.6	123.2	447.3	564.4
1952	1,157.3	124.7	479.3	555.2
1953	1,205.8	114.5	488.1	560.5
1954	1,230.6	118.3	526.3	522.3
1955	1,334.2	116.8	581.1	557.1
1956	1,373.1	117.3	615.9	562.8
1957	1,393.0	107.7	616.1	548.8
1958	1,435.9	113.3	653.4	531.1
1959	1,485.3	109.9	660.0	563.5
1960	1,511.7	132.5	685.3	559.4
1961	1,533.2	144.4	694.4	548.7
1962	1,584.7	164.2	732.4	545.8
1963	1,632.1	172.2	747.3	538.9
1964	1,657.9	92.7	750.4	554.6
1965	1,720.2	97.6	775.8	587.0
1966	1,793.4	101.1	797.4	626.4
1967	1,842.7	100.1	818.2	651.9
1968	1,956.0	102.9	874.5	668.2
1969	2,042.5	100.4	900.3	721.9
1970	2,131.3	96.0	927.2	804.3
1971	2,213.2	90.9	971.3	838.0
1972	2,350.7	85.9	1,066.1	925.6
1973	2,452.7	78.9	1,128.7	1,030.2
1974	2,402.4	64.4	1,075.9	963.2
1975	2,450.3	58.0	1,040.6	898.6
1976	2,567.2	61.9	1,146.7	1,025.1
1977	2,633.5	64.0	1,223.3	1,120.9
1978	2,719.5	64.0	1,252.6	1,103.2
1979	2,581.5	68.6	1,208.5	1,031.6
1980	2,420.5	58.0	1,049.0	918.0
1981	2,415.6	46.3	1,032.5	762.0
1982	2,396.1	47.0	974.9	626.5
1983	2,426.5	46.4	981.9	518.6
1984	2,460.6	42.4	1,042.4	499.5

Source: U.S. Department of Commerce, Business Statistics; Survey of Current Business, various issues.

Table B-6: Percentage Yields of Major Refined Petroleum Products from Crude Petroleum in the U.S. for selected years

Year	Gasoline	Kerosene	Distillate fuel oil	Residual fuel oil	Total
1950	43.0	5.6	19.0	20.2	87.8
1955	44.0	4.3	22.0	15.3	85.6
1960	45.2	4.6	22.4	11.2	83.4
1965	44.0	2.8	22.9	8.0	77.7
1970	45.3	2.3	22.4	6.4	76.4
1973	45.6	1.7	22.5	7.7	77.5
1975	46.5	1.2	21.3	9.9	78.9
1977	43.4	1.3	22.6	12.5	79.9
1978	45.4	1.2	21.6	11.2	79.4
1979	43.0	1.3	21.5	11.5	77.3
1980	44.2	1.2	21.0	12.1	78.5

Source: Commodity Yearbook, 1964-1981; Bureau of Mines.

Note: Figures after 1975 represent estimates.

Table B-7: Wholesale (Producer) Prices of Crude Petroleum and Refined Petroleum in the U.S., 1947 to 1984

(Dollars per barrel)

	CDPRI	RDPRI			
		Refined Petroleum Products			
Year	Crude Petroleum	Weighted average	Gasoline (Regular)	Kerosene	Distillate Fuel oil	Residual Fuel Oil
1947	1.93	2.94	3.59	2.81	2.59	2.17
1948	2.61	3.77	4.35	3.76	3.58	2.89
1949	2.57	3.38	4.42	3.23	3.05	1.72
1950	2.57	3.51	4.49	3.42	3.18	2.00
1951	2.58	3.77	4.72	3.60	3.43	2.24
1952	2.58	3.75	4.72	3.67	3.46	2.01
1953	2.77	3.90	4.95	3.66	3.57	1.97
1954	2.85	3.86	4.72	3.72	3.62	2.11
1955	2.86	3.94	4.72	3.80	3.71	2.37
1956	2.87	4.14	4.85	4.03	3.88	2.69
1957	3.16	4.43	5.08	4.14	4.10	3.20
1958	3.17	4.04	4.75	3.84	3.75	2.52
1959	3.07	4.01	4.73	3.98	3.81	2.37
1960	3.05	4.02	4.77	3.92	3.59	2.53
1961	3.06	4.09	4.77	4.09	3.77	2.61
1962	3.06	4.06	4.72	4.07	3.72	2.57
1963	3.05	4.03	4.67	4.06	3.73	2.48
1964	3.04	3.82	4.47	3.72	3.43	2.41
1965	3.04	3.96	4.60	3.95	3.65	2.48
1966	3.06	4.10	4.86	4.07	3.72	2.42
1967	3.09	4.18	4.92	4.37	3.97	2.30
1968	3.12	4.08	4.72	4.41	4.04	2.20
1969	3.25	4.11	4.82	4.37	4.06	2.15
1970	3.28	4.37	4.99	4.47	4.23	2.89
1971	3.50	4.60	4.99	4.62	4.37	3.82
1972	3.52	4.64	5.12	4.66	4.42	3.66
1973	3.89	5.59	6.12	5.59	5.54	4.38
1974	6.55	10.41	9.94	9.90	10.80	11.18
1975	7.59	11.84	11.80	12.47	12.28	11.41
1976	7.84	12.55	13.01	13.64	13.38	10.43
1977	8.47	13.95	14.12	15.66	15.25	12.03
1978	9.28	14.34	14.76	17.15	15.80	11.47
1979	11.64	20.09	20.47	23.57	22.78	15.76
1980	17.20	30.52	32.12	37.71	33.76	22.14
1981	24.83	36.84	37.09	45.42	42.00	28.53
1982	22.67	34.62	34.11	43.52	40.19	27.22
1983	21.06	31.14	30.73	39.58	35.32	24.39
1984	20.72	30.05	28.71	38.00	34.94	25.80

Source: U.S. Department of Labor, Bureau of Labor Statistics: Wholesale Prices and Price Indexes.

Table B-8: Real Prices* of Crude Petroleum and Refined Petroleum Products in the U.S., 1947 to 1984

Year	Implicit GNP Deflator	Real Price of Crude Petroleum	Real Price of Refined Petroleum Products
1947	49.55	3.90	5.93
1948	52.98	4.92	7.12
1949	52.49	4.90	6.44
1950	53.56	4.80	6.55
1951	57.09	4.53	6.61
1952	57.92	4.46	6.48
1953	58.82	4.71	6.63
1954	59.55	4.79	6.49
1955	60.84	4.69	6.48
1956	62.79	4.57	6.59
1957	64.93	4.87	6.82
1958	66.04	4.80	6.12
1959	67.06	4.57	5.99
1960	68.70	4.44	5.85
1961	69.33	4.41	5.90
1962	70.61	4.34	5.74
1963	71.67	4.26	5.62
1964	72.77	4.18	5.25
1965	74.36	4.08	5.32
1966	76.76	3.98	5.34
1967	79.06	3.91	5.28
1968	82.54	3.77	4.95
1969	86.79	3.75	4.74
1970	91.45	3.59	4.78
1971	96.01	3.64	4.79
1972	100.00	3.52	4.64
1973	105.75	3.68	5.29
1974	115.08	5.69	9.04
1975	125.79	6.04	9.41
1976	132.34	5.92	9.49
1977	140.05	6.05	9.96
1978	150.42	6.17	9.53
1979	163.42	7.12	12.30
1980	178.42	9.64	17.10
1981	195.14	12.73	18.88
1982	206.88	10.96	16.73
1983	215.34	9.78	14.46
1984	223.43	9.28	13.45

*Real prices are nominal prices deflated by the Implicit GNP deflator.

Table B-9: Representative Weights of Petroleum Products in International Trade

Product	Gallons per pound	Pounds per gallon	Pounds per barrel	Barrels per short ton*	Barrels per metric ton*	Barrels per long ton*
Crude Petroleum (Domestic)	0.142	7.03	295	6.77	7.46	7.58
Crude Petroleum (Foreign)	0.133	7.50	315	6.35	7.00	7.11
Gasoline & Naphtha	0.162	6.17	259	7.72	8.51	8.65
Kerosene	0.148	6.75	284	7.05	7.78	7.90
Distillate Fuel Oil	0.138	7.24	304	6.58	7.25	7.37
Residual Fuel Oil	0.127	7.88	331	6.04	6.66	6.77
Lubricating oils	0.133	7.50	315	6.35	7.00	7.11
Paraffin wax	0.150	6.68	280	7.13	7.87	7.99
Petrolatum	-	-	280	7.14	7.87	-
Grease	-	-	350	5.71	6.30	-
Asphalt & road oil	-	-	364	5.50	6.06	-
Petroleum coke	0.105	9.54	401	4.99	5.50	5.59

* 1 short ton = 1 metric ton = 2,000 pounds; 1 long ton = 2,240 pounds.

Sources: Guthrie, V. Petroleum Products Handbook, McGraw Hill Co., 1960; American Petroleum Institute: Petroleum Facts and Figures, 1971.

Note: The above figures are approximate or representative estimates, only, used by the Bureau of Mines in converting international data and should be used only for rough estimating.

Table B-10: Estimated Volume and Value of U.S. General Imports of Crude Petroleum, 1947 to 1984

Year	CGIM Quantity (Millions of barrels)	Value (Millions of dollars) f.a.s.	Value (Millions of dollars) c.i.f.	CFPR Foreign Price ($ per barrel) c.i.f. value/quant
1947	104.10	171.02	185.16	1.78
1948	128.77	286.52	310.21	2.41
1949	159.24	348.62	377.45	2.37
1950	180.25	379.36	410.74	2.28
1951	182.74	385.48	417.36	2.28
1952	210.66	444.90	481.69	2.29
1953	235.02	505.69	547.51	2.33
1954	250.34	561.28	607.70	2.43
1955	300.92	677.73	733.78	2.44
1956	355.01	841.25	910.82	2.57
1957	387.82	986.14	1,067.70	2.75
1958	404.23	995.99	1,078.36	2.67
1959	404.81	931.51	1,008.55	2.49
1960	421.13	951.16	1,029.82	2.45
1961	438.53	1,005.93	1,089.12	2.48
1962	465.92	1,050.48	1,137.36	2.44
1963	470.33	1,065.11	1,153.19	2.45
1964	482.17	1,079.78	1,169.08	2.42
1965	501.49	1,118.23	1,210.71	2.41
1966	496.33	1,115.26	1,207.49	2.43
1967	476.00	1,078.54	1,167.74	2.45
1968	537.36	1,202.87	1,302.35	2.42
1969	579.17	1,317.54	1,426.50	2.46
1970	545.28	1,280.83	1,386.76	2.54
1971	676.46	1,703.73	1,844.63	2.73
1972	901.07	2,383.21	2,580.30	2.86
1973	1,293.71	4,230.81	4,580.71	3.54
1974	1,367.32	15,335.30	16,603.56	12.14
1975	1,584.73	18,374.39	19,754.14	12.47
1976	2,050.42	25,479.88	27,462.40	13.39
1977	2,519.80	33,582.57	35,719.93	14.18
1978	2,392.35	32,140.39	34,263.85	14.32
1979	2,467.32	46,099.98	49,023.68	19.87
1980	1,977.25	62,014.32	64,633.03	32.69
1981	1,763.07	61,940.27	64,318.87	36.48
1982	1,420.75	45,861.86	47,445.45	33.39
1983	1,293.82	36,809.13	38,183.99	29.51
1984	1,319.73	36,528.82	37,945.02	28.75

Notes:
1. These import data were initially compiled in terms of the commodity classification in the Tariff Schedule of the United States Annotated (TSUSA). Starting February 1984, the TSUSA data are presented in terms of the Standard International Classification (SITC).
2. Estimates of quantity and value include unfinished oils reported in the TSUSA classification (1947-1963).
3. C.i.f. values prior to 1974 represent extrapolated values computed on the basis of the corresponding f.a.s. values.
4. All figures from 1954 on do not include shipments of less than $250 resulting from the customs simplification Act of 1953.

Table B-11: Estimated Volume and Value of U.S. General Imports of Refined Petroleum Products, 1947 to 1984

Year	RGIM Quantity (Millions of barrels)	Value (Millions of dollars) f.a.s.	Value (Millions of dollars) c.i.f.	RFPR Foreign Price ($ per barrel) c.i.f. value/quant
1947	62.55	89.46	94.78	1.52
1948	61.17	131.24	139.03	2.27
1949	81.90	129.09	136.75	1.67
1950	127.65	208.48	220.87	1.73
1951	127.65	215.78	228.60	1.79
1952	140.38	246.14	260.77	1.86
1953	147.72	255.89	271.09	1.84
1954	141.36	267.03	282.90	2.00
1955	169.31	354.00	375.03	2.21
1956	182.26	427.38	452.77	2.48
1957	203.38	561.75	595.12	2.93
1958	247.34	629.44	666.83	2.70
1959	267.56	597.03	632.50	2.36
1960	265.91	591.92	627.08	2.36
1961	267.69	609.96	646.19	2.41
1962	319.02	717.20	759.81	2.38
1963	325.46	713.04	755.40	2.32
1964	371.10	800.70	848.27	2.29
1965	432.54	942.04	998.00	2.31
1966	473.49	1,002.54	1,062.09	2.24
1967	485.02	1,025.83	1,086.77	2.24
1968	541.32	1,162.21	1,231.25	2.27
1969	605.19	1,263.15	1,338.19	2.21
1970	715.36	1,524.77	1,615.35	2.26
1971	700.62	1,676.91	1,776.53	2.54
1972	782.86	1,989.76	2,107.96	2.69
1973	988.82	3,468.90	3,674.97	3.72
1974	848.81	9,308.73	9,861.72	11.62
1975	605.10	6,814.78	7,212.96	11.92
1976	583.70	6,731.27	7,088.02	12.14
1977	638.20	8,543.46	8,970.53	14.06
1978	570.90	7,397.42	7,777.60	13.62
1979	555.12	10,633.42	11,195.54	20.17
1980	460.82	12,767.20	13,351.41	28.97
1981	466.17	14,802.73	15,413.98	33.07
1982	491.83	14,667.18	15,289.75	31.09
1983	558.51	16,462.27	17,146.01	30.70
1984	685.91	20,453.78	21,328.99	31.10

Notes:
1. Same note as in Table 11.
2. Quantity and value estimates include partly refined products reported in Schedule A, SITC classification.
3. C.i.f. values prior to 1974 represent extrapolated values computed on the basis of the corresponding f.a.s. values.
4. Same note as in Table 11.

Table B-12: F.a.s. Prices of Imported Crude Petroleum and Products

(Dollars per barrel)

Year	Crude Petroleum	Refined Products	Gasoline	Kerosene	Distillate fuel oil	Residual fuel oil
1947	1.64	1.43	4.75	3.53	1.84	1.38
1948	2.23	2.15	4.25	3.78	2.75	2.12
1949	2.19	1.58	4.11	10.63	3.49	1.59
1950	2.10	1.63	4.45	2.74	2.28	1.60
1951	2.11	1.69	4.79	4.30	2.76	1.61
1952	2.11	1.75	4.96	3.51	2.89	1.65
1953	2.15	1.73	4.99	6.60	2.82	1.64
1954	2.24	1.89	5.01	9.58	3.16	1.77
1955	2.25	2.09	5.19	3.75	3.08	1.89
1956	2.37	2.34	4.57	3.87	3.23	2.15
1957	2.54	2.76	4.21	4.29	3.42	2.56
1958	2.46	2.54	3.88	4.40	3.21	2.25
1959	2.30	2.23	3.52	4.30	3.48	1.97
1960	2.26	2.23	4.24	3.22	3.06	2.05
1961	2.29	2.28	3.88	3.59	3.03	2.12
1962	2.25	2.25	3.31	2.66	2.69	2.12
1963	2.26	2.19	3.26	4.15	2.58	2.32
1964	2.24	2.16	3.66	3.42	2.69	2.04
1965	2.23	2.18	3.38	3.55	2.76	2.05
1966	2.25	2.12	3.35	3.71	2.24	1.98
1967	2.27	2.12	3.63	2.83	2.42	1.98
1968	2.24	2.15	3.82	2.99	2.44	2.35
1969	2.27	2.09	3.97	2.66	2.67	2.16
1970	2.35	2.13	4.14	3.17	2.79	2.22
1971	2.52	2.39	4.25	3.69	2.64	2.41
1972	2.64	2.54	5.10	4.82	2.25	2.35
1973	3.27	3.51	7.87	6.45	3.29	3.48
1974	11.22	10.97	16.66	19.84	11.63	10.92
1975	11.59	11.26	13.75	11.62	12.06	10.37
1976	12.43	11.53	15.88	0.00	11.78	11.36
1977	13.33	13.39	15.99	15.74	14.33	13.25
1978	13.43	12.96	16.75	11.77	13.41	12.63
1979	18.68	19.16	32.25	21.01	28.75	17.83
1980	31.36	27.71	37.05	26.48	32.93	27.74
1981	35.13	31.75	39.39	40.04	35.37	31.77
1982	32.28	29.82	36.66	39.31	33.48	29.79
1983	28.45	29.48	33.37	33.29	30.70	28.52
1984	27.68	29.82	31.20	34.06	30.90	28.65

Notes:
1. Prices of crude petroleum and refined products represent the weighted price computed by dividing the total f.a.s. value by the corresponding quantity (see Table 11).
2. Prices of gasolene, kerosene, distillate fuel oil and residual fuel oil represent simple averages.

Table B-13: Estimates of Regular Grade Gasoline Prices, State and Federal Taxes and Refined Products Composite Price

(dollars per barrel)

Year	Regular Grade Gasoline		Refined Products Composite Price	
	Gasoline Price Incl. Taxes	State and Federal Taxes	REFPRI Current Dollars	REFPRID Constant 1972 dollars
1947	9.71	2.60	5.77	11.65
1948	10.87	2.66	6.88	12.99
1949	11.25	2.74	6.77	12.90
1950	11.24	2.81	6.76	12.63
1951	11.40	2.87	7.04	12.34
1952	11.49	3.07	7.14	12.32
1953	12.05	3.11	7.51	12.77
1954	12.20	3.14	7.70	12.93
1955	12.21	3.21	7.81	12.83
1956	12.57	3.51	8.11	12.91
1957	13.00	3.72	8.57	13.19
1958	12.76	3.74	8.25	12.49
1959	12.81	3.91	8.27	12.33
1960	13.07	4.26	8.36	12.17
1961	12.92	4.30	8.37	12.07
1962	12.87	4.32	8.32	11.78
1963	12.78	4.33	8.31	11.59
1964	12.75	4.36	8.31	11.42
1965	13.08	4.39	8.54	11.49
1966	13.47	4.41	8.76	11.41
1967	13.93	4.46	9.04	11.43
1968	14.16	4.53	9.21	11.15
1969	14.63	4.62	9.44	10.87
1970	14.99	4.68	9.76	10.67
1971	15.30	4.72	10.14	10.57
1972	15.17	4.90	9.97	9.97
1973	16.30	5.01	10.92	10.32
1974	22.01	5.04	16.85	14.64
1975	24.33	5.22	18.75	14.90
1976	25.35	5.44	19.15	14.47
1977	27.11	5.82	20.73	14.81
1978	28.39	6.09	21.55	14.33
1979	38.93	8.35	29.83	18.26
1980	56.46	12.11	43.77	24.53
1981	62.56	13.42	51.30	26.29
1982	57.85	12.41	48.69	23.53
1983	51.87	11.13	44.05	20.46
1984	49.19	10.55	42.51	19.03

Table B-14: Estimates of Federal Tax Rates and Federal Regulations

Year	Crude Petroleum			Refined Petroleum Product		
	FTAXC (%)	TAXC ($ per barrel)	REGC ($ per barrel)	FTAXR (%)	TAXR ($ per barrel)	REGR ($ per barrel)
1947	6.61	0.26	-0.31	2.82	0.33	-2.87
1948	6.66	0.33	-0.37	3.34	0.43	-2.83
1949	5.50	0.27	-0.38	1.96	0.25	-3.26
1950	6.24	0.30	-0.55	3.08	0.39	-3.32
1951	7.37	0.33	-0.53	4.19	0.52	-3.47
1952	7.91	0.35	-0.51	2.70	0.33	-3.27
1953	8.08	0.38	-0.75	2.82	0.36	-3.51
1954	7.98	0.38	-0.71	2.46	0.32	-3.12
1955	7.26	0.34	-0.69	2.70	0.35	-2.84
1956	7.16	0.33	-0.49	2.68	0.35	-2.63
1957	6.94	0.34	-0.63	1.77	0.23	-2.31
1958	6.57	0.32	-0.76	1.40	0.17	-2.04
1959	6.31	0.29	-0.86	1.73	0.21	-2.46
1960	6.45	0.29	-0.88	1.60	0.19	-2.42
1961	7.58	0.33	-0.83	1.75	0.21	-2.42
1962	8.26	0.36	-0.88	1.77	0.21	-2.37
1963	7.73	0.33	-0.84	2.34	0.27	-2.38
1964	9.03	0.38	-0.84	2.12	0.24	-2.10
1965	8.89	0.36	-0.83	2.34	0.27	-2.22
1966	9.38	0.37	-0.81	2.73	0.31	-2.42
1967	9.78	0.38	-0.81	2.80	0.32	-2.45
1968	9.69	0.37	-0.84	2.78	0.31	-2.19
1969	8.88	0.33	-0.91	2.54	0.28	-2.19
1970	10.05	0.36	-0.80	2.55	0.27	-2.31
1971	12.02	0.44	-0.80	2.86	0.30	-2.15
1972	14.21	0.50	-0.65	2.41	0.24	-1.95
1973	19.41	0.71	-0.33	3.00	0.31	-1.77
1974	25.72	1.46	4.86	2.67	0.39	1.05
1975	24.30	1.47	3.87	2.94	0.44	0.06
1976	22.96	1.36	4.20	2.64	0.38	-0.31
1977	20.27	1.23	4.07	2.42	0.36	0.08
1978	19.78	1.22	3.36	2.46	0.35	-0.47
1979	19.84	1.41	5.04	3.12	0.57	0.05
1980	24.46	2.36	8.68	3.43	0.84	-0.87
1981	23.87	3.04	5.97	2.58	0.68	-1.93
1982	18.60	2.04	5.19	2.33	0.55	-1.71
1983	14.49	1.42	3.93	2.10	0.43	-0.20
1984	11.29	1.05	3.59	1.89	0.36	0.47

Notes: The labels of this table are defined as follows:
FTAXC = Corporation Income Tax Rate as a percent of Gross Receipts in the Oil and Gas Extraction Industry
TAXC = The federal tax in dollars per barrel of crude petroleum
REGC = The difference between CFPRID and CDPRID
FTAXR = Corporation Income Tax Rate as a percent of Gross Receipts in the Petroleum and Coal Products Industry
TAXR = The federal tax in dollars per berrel of refined petroleum products
REGR = The difference between RFPRID and RDPRID

Table B-15: Estimates of Crude Petroleum and Refined Petroleum Products Taxes

(dollars per barrel)

Year	QC	QR
1947	-0.0569	5.5670
1948	-0.0431	5.4600
1949	-0.1137	5.4698
1950	-0.2470	5.6272
1951	-0.1917	5.5490
1952	-0.1602	5.6406
1953	-0.3719	5.6513
1954	-0.3318	5.5936
1955	-0.3452	5.6275
1956	-0.1593	5.9380
1957	-0.2959	5.9582
1958	-0.4468	5.8414
1959	-0.5683	6.0443
1960	-0.5902	6.3939
1961	-0.4925	6.4086
1962	-0.5223	6.3233
1963	-0.5062	6.3131
1964	-0.4661	6.2273
1965	-0.4721	6.1712
1966	-0.4392	6.0621
1967	-0.4240	5.9566
1968	-0.4724	5.7954
1969	-0.5757	5.5946
1970	-0.4445	5.3883
1971	-0.3658	5.2148
1972	-0.1538	5.1418
1973	0.3805	5.0518
1974	6.3266	4.7704
1975	5.3397	4.5884
1976	5.5578	4.4919
1977	5.2972	4.5118
1978	4.5750	4.4011
1979	6.4505	5.6789
1980	11.0403	7.6304
1981	9.0065	7.5570
1982	7.2233	6.5475
1983	5.3423	5.5981
1984	4.6407	5.0835

Note: See the glossary of symbols in Appendix B for the definition of QC and QR.

Table B-16: Laws of Motion of the Exogenous State Variables

Variable	QC	QR	PS	W	X	Y
Constant*					0.06 (0.01)	83.79 (35.80)
D1**						
D2						
D3	4.53 (0.77)					
Time						6.78 (3.19)
Lagged Dep. Variable	0.31 (0.12)	1.00 (0.00)	.90 (.07)	1.09 (0.04)		0.80 (0.10)
Statistics						
Mean	1.68	5.79	9.2	0.49	0.06	994.04
Standard Deviation	3.31	0.39	3.19	0.17	0.03	349.10
Stand. Error of the Reg.	1.02	0.14	1.06	0.07	0.03	27.21
$\tilde{R}^2$	0.91	0.86	0.89	0.75	0.00	0.99
Durbin-Watson	1.29	1.16	1.35	1.52	1.86	1.58
F	345.47					2945.8

Notes:

* No entry means that the variable was found insignificant and dropped.

**

$D_1 = \begin{cases} 1, & 1947\text{-}1958 \\ 0, & \text{otherwise} \end{cases}$ $D_2 = \begin{cases} 1, & 1959\text{-}1973 \\ & \text{(period of quotas)} \\ 0, & \text{otherwise} \end{cases}$ $D_3 = \begin{cases} 1, & 1974\text{-}1984 \\ & \text{(period of price controls)} \\ 0, & \text{otherwise} \end{cases}$

References

Adams, Gerard F. and James M.Griffin (1972): "An Econometric-Linear Programming Model of the U.S. Petroleum Refining Industry," Journal of the American Statistical Association, 67, 542-551.

Adelman, Morris A. (1972): The World Petroleum Market, Washington, DC: Resources for the Future.

Aiyagari, Rao S. and Raymond Riezman (1985): "Analysis of Embargoes and Supply Shocks in a Market with a Dominant Seller," in Energy, Foresight and Strategy, ed. by Thomas J. Sargent, Washington, DC: Resources for the Future.

Blanchard, Olivier J. (1983): "The Production and Inventory of the American Automobile Industry," Journal of Political Economy, 91, 365-400.

Blinder, Alan S. (1984): "Can the Production Smoothing Model of Inventory Behavior be Saved?", Working Paper No. 1257, Cambridge, MA: National Bureau of Economic Research.

Bohi, Douglas and Milton Russel (1978): Limiting Oil Imports, Baltimore: John Hopkins University Press.

Chao, Hung-Po and Alan S. Manne (1982): "An Integrated Analysis of U.S. Stockpiling Policies." in Energy Vulnerability, ed. by J. Plummer, Cambridge, MA: Ballinger.

Eckstein, Otto (1978): The Great Recession, with a Postscript on Stagflation, New York: North-Holland.

Eckstein, Otto. (1981): "Shock Inflation, Core Inflation, and Energy Disturbances in the DRI Model." in Energy Prices, Inflation and Economic Activity, ed. by Knut A. Mork, Cambridge, MA: Ballinger.

Eckstein, Zvi and Martin S. Eichenbaum (1985a): "Oil Supply Disruptions and the Optimal Tariff in a Dynamic Stochastic Equilibrium Model," in Energy, Foresight and Strategy, ed. by Thomas J. Sargent, Washington, DC: Resources for the Future.

Eckstein, Zvi and Martin Eichenbaum (1985b): "Inventories and Quantity-Constrained Equilibria in Regulated Markets: the U.S. Petroleum Industry, 1947-1972," in Energy, Foresight and Strategy, ed. by Thomas J. Sargent, Washington, DC: Resources for the Future.

Eichenbaum, Martin S. (1983): "A Rational Expectations Equilibrium of Inventories of Finished Goods and Employment," Journal of Monetary Economics, 13, (1983): 259-277.

Eichenbaum, Martin S. (1984): "Rational Expectations and the Smoothing Properties of Finished Goods," Journal of Monetary Economics, 14 71-96.

Epple, Dennis (1985): "The Econometrics of Exhaustible Resource Supply: A Theory and an Application," in *Energy, Foresight and Strategy*, ed. by Thomas J. Sargent, Washington, DC: Resources for the Future.

Epple, Dennis and Lars P. Hansen (1981): "An Econometric Framework for Modeling Exhaustible Resource Supply," in *The Economics of Exploration of Energy Resources*, ed. by James B. Ramsey, Greenwich, CH: JAI Press.

Fair, Ray C. (1978): "Inflation and Unemployment in a Macroeconomic Model," in *After the Phillips Curve: Persistence of High Inflation and High Unemployment*, Boston: Federal Reserve Bank of Boston.

Goldfeld, Stephen M. and Richard E. Quandt (1972): *Nonlinear Methods in Econometrics*, Amsterdam: North-Holland.

Hamilton, James D. (1983): "Oil and the Macroeconomy since World War II," *Journal of Political Economy*, 91, 228-248.

Hansen, Lars P., Epple Dennis and W. Roberds (1985): "Linear-Quadratic Duopoly Models Games of Resource Depletion," in *Energy, Foresight and Strategy*, ed. by Thomas J. Sargent, Washington, DC: Resources for the Future.

Hansen, Lars P. and Thomas J. Sargent (1980): "Formulating and Estimating Dynamic Linear Rational Expectations Models," *Journal of Economic Dynamics and Control,* 2, 7-46.

Hansen, Lars P. and Thomas J. Sargent (1981): "Linear Rational Expectations Models for Dynamically Interrelated Variables," in *Rational Expectations and Econometric Practice*, ed. by Robert E. Lucas, Jr. & Thomas J. Sargent, Minneapolis, MN: University of Minnesota Press.

Houthakker, Hendrick S. and M. Kennedy (1979): "A Long-Run Model of World Energy Demands, Supplies and Prices," In *Directions in Energy Policy: A Comprehensive Approach to Energy Resource Decision-Making*, ed. by Behram Kursunoglou and Arnold Perlmutter.

Houthakker, Hendrick S. and Lester Taylor (1970): *Consumer Demand in the United States: Analysis and Projections*, Cambridge, MA: Harvard University Press.

Hubbard, R. Glenn and Robert J. Weiner (1983): "The Sub-trigger" Crisis: An Economic Analysis of Flexible Stock Policies." *Energy Economics*, 178-189.

Hudson, Edward A. and Dale W. Jorgenson (1978): "Energy Prices and the U.S. Economy, 1972-1976," *Natural Resources Journal*, 18, 877-897.

Husted, Steven L. and Tryphon Kollintzas (1982): "Estimation of Dynamic Import Demand Functions Under Quasi-Rational Expectations," in Time Series: Theory & Practice 2, ed. by Oliver D. Anderson, Amsterdam: North-Holland.

Husted, Steven L. and Tryphon Kollintzas (1984): "Import Demand with Rational Expectations: Estimates for Bauxite, Cocoa, Coffee and Petroleum," Review of Economics and Statistics, 66, 608-618.

Husted, Steven L. and Tryphon Kollintzas (1987): "Linear Rational Expectations Equilibrium Laws of Motion for Selected U.S. Raw Material Imports," International Economic Review, 28, 651-670.

Kalt, Joseph P. (1981): The Economics and Politics of Oil Price Regulation: Federal Policy in the Postembargo Era, Cambridge, MA: The MIT Press.

Kang, Heejoon (1985): "The Effects of Detrending in Granger Causality Tests," Journal of Business and Economic Statistics, 3, 344-349.

Kennedy, Michael (1976): "A World Oil Model," in Economic Studies of U.S. Energy Policy, ed. by Dale W. Jorgenson, Amsterdam: North-Holland.

Klein, Lawrence R. (1978): "Disturbances to the International Economy," in After the Phillips Curve: Persistence to High Inflation and High Unemployment, Boston: Federal Reserve Bank of Boston.

Kollintzas, Tryphon (1985): "The Symmetric Linear Rational Expectations Model," Econometrica, 53, 963-976.

Kollintzas, Tryphon and Harrie L. A. M. Geerts (1984): "Three Notes on the Formulation of Linear Rational Expectations Models," in Time Series: Theory and Practice 5, ed. by Oliver D. Anderson, Amsterdam: North-Holland.

Kollintzas, Tryphon and Steven L. Husted (1984): "Distributed Lags and Intermediate Good Imports," Journal of Economic Dynamics and Control, 8, 303-327.

Lovell, Michael (1961): "Manufacturer's Inventories Sales Expectations, and the Acceleration Principle," Econometrica, 29, 293-314.

Lucas, Robert E. Jr. (1976): "Econometric Policy Evaluation: A Critique." In The Phillips Curve and Labor Markets, edited by Karl Brunner & Allan H. Meltzer, Amsterdam: North-Holland.

Lucas, Robert E. Jr. and Edward C. Prescott (1971): "Investment Under Uncertainty," Econometrica, 39, 659-81.

MacAvoy, Paul and Robert Pindyck (1973): "Alternative Regulatory Policies for Dealing with the Natural Gas Shortage," Bell Journal of Economics, 4, 454-498.

Maccini, Louis J. (1984): "The Interrelationship Between Price and Output Decisions and Inventory Decisions, Microfoundations and Aggregate Implications," Journal of Monetary Economics, 13, 41-65.

Mankiw, Gregory and Matthew D. Shapiro (1985): "Trends, Random Walks, and Tests of the Permanent Income Hypothesis," Journal of Monetary Economics, 16, 165-174.

Mitchell, Edward J. (1976): Vertical Integration in the Oil Industry, Washington, DC: American Enterprise Institute for Public Policy Research.

Mork, Knut A. and Robert E. Hall (1980a): "Energy Prices and the U.S. Economy in 1979-1981," The Energy Journal, 1, 41-53.

Mork, Knut A. and Robert E. Hall (1980b): "Energy Prices, Inflation, and Recession, 1974-1975," The Energy Journal, 1(3), 31-63.

Mork, Knut A. and Robert Hall (1981): "Macroeconomic Analysis of Energy Price Shocks and Offsetting Policies: An Integrated Approach," in Energy Prices, Inflation, and Economic Activity, ed. by K. Mork, Cambridge, MA: Ballinger.

Nelson, Charles R. and Heejoon Kang (1981): "Spurious Periodicity in Inappropriately Detrended Time Series," Econometrica, 49, 741-751.

Nelson, Charles R. and Heejoon Kang (1984): "Pitfalls in the Use of Time as an Explanatory Variable in Regression," Journal of Business and Economic Statistics ,2, 73-82.

Nelson, Charles R. and Charles I. Plosser (1982): "Trends and Random Walks in Macroeconomic Time Series: Some Evidence and Implications," Journal of Monetary Economics, 10, 139-162.

Nerlove, Marc (1972): "Lags in Economic Behavior," Econometrica, 40, 221-245.

Nordhaus, William D. (1973): "The Allocation of Energy Resources," Brookings Papers in Economic Activity, 3, 529-576.

Perry, George L. (1975): "Policy Alternatives for 1974," Brookings Papers on Economic Activity, 1, 222-237.

Pierce, James L. and Jared J. Enzler (1974): "The Effects of External Inflationary Shocks," Brookings Papers on Economic Activity, 1, 13-61.

Rice, Patricia and Kerry V. Smith (1977): "An Econometric Model of the Petroleum Industry," Journal of Econometrics, 6, 263-287.

Sargent, Thomas J. (1979): *Macroeconomic Theory*, New York: Academic Press.

Sargent, Thomas J. (1981): "Interpreting Economic Time Series," *Journal of Political Economics*, 89, 213-248.

Sargent, Thomas J. (1984): "Autoregressions, Expectations and Advice," *American Economic Review*, 74, 408-420.

Sargent, Thomas J. (1985): *Energy, Foresight and Strategy*, Washington, DC: Resources for the Future.

Stokey, Nancy L., Robert E. Lucas, and Edward C. Prescott (1987): "Recursive Methodss in Economic Dynamics," Manuscript, Northwestern University.

Verleger, Philip K. (1979): "The U.S. Petroleum Crisis of 1979," *Brookings Papers on Economic Activity*, 2, 463-476.

Verleger, Philip K. (1982a): *Oil Markets in Turmoil*, Cambridge, MA: Ballinger.

Verleger, Philip K. (1982b): "The Determinants of Official OPEC Crude Prices," *Review of Economics and Statistics*, 64, 177-183.

West, Kenneth D. (1986): "A Variance Bounds Test of the Linear Quadratic Inventory Model," *Journal of Political Economy*, 94, 375-401.

CHAPTER V

SEASONALITY, COST SHOCKS, AND THE
PRODUCTION SMOOTHING MODEL OF INVENTORIES[1]

Jeffrey A. Miron
University of Michigan and NBER

Stephen P. Zeldes
The Wharton School, University of Pennsylvania and NBER

Abstsract

A great deal of research on the empirical behavior of inventories examines some variant of the production smoothing model of finished goods inventories. The overall assessment of this model that exists in the literature is quite negative: there is little evidence that manufacturers hold inventories of finished goods in order to smooth production patterns.

This paper examines whether this negative assessment of the model is due to one or both of two features: cost shocks and seasonal fluctuations. The reason for considering cost shocks is that, if firms are buffeted more by cost shocks than demand shocks, production should optimally be more variable than sales. The reasons for considering seasonal fluctuations are that seasonal fluctuations account for a major portion of the variance in production and sales, that seasonal fluctuations are precisely the kinds of fluctuations that producers should most easily smooth, and that seasonally adjusted data are likely to produce spurious rejections of the production smoothing model even when it is correct.

We integrate cost shocks and seasonal fluctuations into the analysis of the production smoothing model in three steps. First, we present a general production smoothing model of inventory investment that is consistent with both seasonal and non-seasonal fluctuations in production, sales, and inventories. The model allows for both

---

[1] The authors are grateful for helpful comments by Andrew Abel, Oliver Blanchard, Alan Blinder, John Campbell, Larry Christiano, Angus Deaton, Martin Eichenbaum, Stanley Fischer, Marvin Goodfriend, Tryphon Kollintzas, Louis Maccini, Trish Mosser, Chris Sims, Larry Summers, Ken West, members of the Macro Lunch Group at the University of Pennsylvania, two referees, and seminar participants at workshops at Michigan, Princeton, Johns Hopkins, Rochester, McMaster, Northwestern, Econometric Society/NORC, and FRB Minneapolis and Philadelphia; and for excellent research assistance by Edward Gold. This chapter is reprinted, with permission, from Econometrica, 56, July 1988, pp. 877-908.

observable and unobservable changes in marginal costs. Second, we estimate this model using both seasonally adjusted and seasonally unadjusted data plus seasonal dummies. The goal here is to determine whether the incorrect use of seasonally adjusted data has been responsible for the rejections of the production smoothing model reported in previous studies. The third part of our approach is to explicitly examine the seasonal movements in the data. We test whether the residual from an Euler equation is uncorrelated with the seasonal component of contemporaneous sales. Even if unobservable seasonal cost shocks make the seasonal variation in output greater than that in sales, the timing of the resulting seasonal movements in output should not necessarily match that of sales.

The results of our empirical work provide a strong negative report on the production smoothing model, even when it includes cost shocks and seasonal fluctuations. At both seasonal and non-seasonal frequencies, there appears to be little evidence that firms hold inventories in order to smooth production. A striking piece of evidence is that in most industries the seasonal in production closely matches the seasonal in shipments, even after accounting for the movements in interest rates, input prices, and the weather.

1. Introduction

A great deal of research on the empirical behavior of inventories examines some variant of the production smoothing model of finished goods inventories. Blinder (1986a) emphasizes that, in the absence of cost shocks, the model implies that the variance of production should be less than the variance of sales, an inequality that is violated for manufacturing as a whole and most 2-digit industries. West (1986) derives a variance bounds test that extends this inequality in a number of ways and also finds that the data reject the model. Both Blinder and West conclude that there is strong evidence against the production smoothing model. Other authors, such as Blanchard (1983), Eichenbaum (1984), and Christiano and Eichenbaum (1986), present evidence that is less unfavorable to the model, but they reject it as well.

This paper examines the extent to which the negative assessment of the model is due to two features: cost shocks and seasonal fluctuations. Blinder (1986a) and West (1986) both note that the presence of cost shocks could explain the rejections that they report, and Blinder (1986b), Maccini and Rossana (1984), Eichenbaum (1984), and Christiano and Eichenbaum (1986) test the model in the presence of cost shocks, with partial success. The reason for considering these shocks is simply that if firms are buffeted more by cost shocks than demand shocks, production should optimally be more variable than sales.

Most of the empirical work on the production smoothing model uses data adjusted by the X-11 seasonal adjustment routine. This includes studies by Blinder (1986a, 1986b), Eichenbaum (1984), and Maccini and Rossana (1984). Blanchard (1983), Reagan and Sheehan (1985), and West (1986) begin with the seasonally unadjusted data and then adjust the data with seasonal dummies. Few studies examine whether the seasonal fluctuations themselves are consistent with the model of inventories. Exceptions are Ward (1978), who finds evidence that firms alter production rates differently in response to seasonal versus nonseasonal variations in demand; West (1986), who includes a version of his variance bounds test based on both the seasonal and nonseasonal variations in the data; and Ghali (1987), who uses data from the Portland Cement industry and finds that seasonal adjustment of the data is an important factor in the rejection of the production smoothing model.[2]

There are several reasons to think that using seasonally adjusted data to test inventory models is problematic. To begin with, seasonal fluctuations account for a major portion of the variation in production, shipments, and inventories. Table 1 shows the seasonal, non-seasonal, and total variance of the logarithmic rate of growth of production and shipments, for six 2-digit manufacturing industries.[3,4] For both variables, seasonal variation accounts for more than half of the total variance in most industries. Any analysis of production/inventory behavior that excludes seasonality at best explains only part of the story and fails to exploit much of the variation in the data.

Seasonal fluctuations are likely to be particularly useful in examining the production smoothing model because they are anticipated.

---

[2] Irvine (1981) uses seasonally unadjusted data, with no seasonal dummies, to examine retail inventory behavior and the cost of capital.

[3] Table 1 includes results based on two different measures of production. See Section 4 for details.

[4] This table is similar to Table 2 in Blanchard (1983). As he points out, since the seasonal component is deterministic, it has no variance in the statistical sense. The numbers reported here for the seasonal variances are the average squared deviations of the twelve seasonal dummy coefficients from the sample mean of these coefficients.

Any test of the production smoothing model involves a set of maintained hypotheses, one of which is the rationality hypothesis. Rejections of the model, therefore, are not usually informative as to which aspect of the joint hypothesis has been rejected. When a rational expectations model is applied to seasonal fluctuations, however, it seems reasonable to take the rationality hypothesis as correct, since if anything is correctly anticipated by agents seasonal fluctuations ought to be. Applying the production smoothing model to seasonal fluctuations may help determine which aspects of the model, if any, fail.

A final reason to avoid the use of seasonally adjusted data is that, since the true model must apply to the seasonally unadjusted data, the use of adjusted data is likely to lead to rejection of the model even when it is correct.[5] This is especially the case with data adjusted by the Census X-11 method because this technique makes the adjusted data a two-sided moving average of the underlying unadjusted data.[6,7] Therefore, the key implication of most rational expectations models, that the error term should be uncorrelated with lagged information, need not hold in the adjusted data even if it does hold in the unadjusted data.[8] If the data are adjusted by some other method, such as seasonal dummies, then the time series properties of the adjusted data are not altered as radically as they are with X-11.

We integrate cost shocks and seasonal fluctuations into the analysis of the production smoothing model in three steps. First, we present a general production smoothing model of inventory investment that is consistent with both seasonal and non-seasonal fluctuations in production, sales, and inventories. The model allows for both observable and unobservable changes in marginal costs (cost shocks).

---

[5] Summers (1981) emphasizes this point.

[6] X-11 is not literally a two-sided moving average filter. Rather, it can be well approximated by such filters. For more on this point, see Cleveland and Tiao (1976) and Wallis (1974).

[7] For example, Miron (1986) finds that the use of X-11 adjusted data is partially responsible for rejections of consumption Euler equations.

[8] See Sargent (1978).

Table 1

Summary Statistics, Log Growth Rates of Production and Shipments

	Food	Tobacco	Apparel	Chemicals	Petroleum	Rubber
Shipments						
Mean:	0.16%	-0.09%	-0.06%	0.20%	0.16%	0.15%
Variance:						
Nonseasonal	5.60E-04	5.08E-03	1.74E-03	9.30E-04	5.10E-04	1.08E-03
Seasonal	1.63E-03	1.61E-03	1.36E-02	2.86E-03	3.10E-04	4.33E-03
Total	2.19E-03	6.69E-03	1.54E-02	3.79E-03	8.20E-04	5.41E-03
Seasonal/Total	75%	24%	89%	75%	38%	80%
Production, Y4						
Mean:	0.13%	-0.17%	-0.04%	0.19%	0.13%	0.15%
Variance:						
Nonseasonal	6.10E-04	6.03E-03	3.10E-03	8.00E-04	9.80E-04	1.60E-03
Seasonal	1.63E-03	5.26E-03	1.16E-02	2.20E-03	2.20E-04	5.13E-03
Total	2.24E-03	1.13E-02	1.47E-02	3.00E-03	1.20E-03	6.73E-03
Seasonal/Total	73%	47%	79%	73%	18%	76%
Production, IP						
Mean:	0.22%	-0.17%	0.01%	0.36%	0.11%	0.42%
Variance:						
Nonseasonal	1.50E-04	2.47E-03	1.80E-03	2.70E-04	4.50E-04	1.09E-03
Seasonal	8.50E-04	1.70E-02	4.28E-03	4.30E-04	5.10E-04	3.06E-03
Total	1.00E-03	1.95E-02	6.08E-03	6.90E-04	9.60E-04	4.14E-03
Seasonal/Total	85%	87%	70%	62%	53%	74%

Notes: The sample period is 1967:5-1982:12. The log growth rates are defined as $\ln x_t - \ln x_{t-1}$. They have not been annualized.

The observables include wages, energy prices, raw materials prices, and interest rates, as well as weather variables (temperature and precipitation).[9] We examine a firm's cost minimization problem, so our analysis is robust to various assumptions about the competitiveness of the firm's output market. A key implication of the model is that, for any firm that can hold finished goods inventories at finite cost, the marginal cost of producing an additional unit of output today and holding it in inventories until next period must equal the expected marginal cost of producing that unit next period. With standard types of auxiliary assumptions about functional forms and identification, the model leads to an estimable Euler equation relating the rate of growth of production to the rate of growth of input prices, the level of inventories, and the interest rate. We estimate this Euler equation and test the overidentifying restrictions implied by the model, using data on six 2-digit manufacturing industries.

The second part of our approach is to perform the exact same estimations and tests of the above model using both seasonally adjusted and unadjusted data.[10] The goal here is to determine whether the incorrect use of seasonally adjusted data is responsible for the rejections of the production smoothing model reported in previous studies.

The third part of our approach is to explicitly examine the seasonal movements in the data. Since the predictable seasonal movement in demand is exactly the variation that should be most easily smoothed by firms, tests of the model at seasonal frequencies are particularly powerful. We therefore test whether the residual from the Euler equation is uncorrelated with the seasonal component of contemporaneous sales. Even if unobservable seasonal cost shocks make

---

[9] Maccini and Rossana (1984) estimate a different style model of inventory accumulation (a general flexible accelerator model) using data on aggregate durables and non-durables inventory accumulation in which they include wages, energy costs, interest rates, and raw materials prices. They found that only raw materials prices had significant effects in their model.

[10] Constant dollar, seasonally unadjusted inventory data are not available and therefore constructed. This is discussed further in Section 4.

the seasonal variation in output greater than that of sales, the <u>timing</u> of the resulting seasonal movements in output should not match that of sales.

The estimation strategy that we employ involves a number of important identifying restrictions about the shifts over time in the firm's production function. We include a number of observable variables that account for the shifts in technology. There may, however, be additional shifts that are not accounted for by the measured variables, and these unobserved productivity shifts will appear in the error term of the equation that we estimate. In order to consistently estimate the Euler equation, therefore, we need to assume that this term is uncorrelated with the variables we use as instruments. Specifically, we assume that the unobserved productivity shifter is uncorrelated with lagged values of sales and with the part of current sales that is predictable based on lagged information, and that the growth of the unobserved productivity shifter is uncorrelated with lagged growth rates of input prices, lagged growth rates of output, and lagged interest rates. In addition, when we examine the seasonal fluctuations in production, we assume that any seasonal in unobserved productivity is uncorrelated with the seasonal in sales.

The remainder of the paper is organized as follows. Section 2 presents the basic production smoothing model that we employ throughout the paper and derives the first order condition that we estimate. In Section 3 we describe the identifying assumptions, the resulting testable implications, and the econometric techniques used to test those implications. In Section 4, we discuss the data used. Section 5 presents the basic results with seasonally adjusted and unadjusted data. In Section 6, we examine the seasonal-specific results. Section 7 concludes the paper.

2. **The Model**

Consider a profit maximizing firm. Sales by the firm, the price of the firm's output, and the firm's capital stock may be exogenously or endogenously determined. The firm may be a monopolist, a perfect competitor, or something in between. The firm is, however, assumed to be a competitor in the markets for inputs. For any pattern of prices, sales, and the capital stock, the firm chooses its input over time so as to minimize costs.

The firm's intertemporal cost minimization problem is

$$\min_{\{y_{t+j}\}} E_t \sum_{j=0}^{T} \Gamma_{t,t+j} C_{t+j}(y_{t+j}) \qquad (1)$$

subject to

$$n_{t+j} = n_{t+j-1}(1 - s'_{t+j-1}) + y_{t+j} - x_{t+j},$$

$$n_{t+j} \geq 0 \quad \forall j,$$

where y_t is production in period t, x_t is sales in period t, and n_t is the stock of inventories at the end of period t, all measured in terms of the output good. The end of the firm's horizon is period T. C_t is the one period nominal cost function of the firm, to be derived shortly. $\Gamma_{t,t+j}$ is the nominal discount factor, defined as the present value at time t of one dollar at t+j. Thus,

$$\Gamma_{t,t+j} = [\prod_{s=0}^{j-1} (\frac{1}{1 + \tilde{R}_{t+s}})], \quad \Gamma_{t,t} = 1,$$

$$\text{and } \tilde{R}_{t+s} = (1 - m_{t+s+1})R_{t+s}.$$

R_t is the pretax cost of capital for the firm, and m_t is the marginal tax rate. E_t indicates expectations conditional on information available at time t.

The term s'_t is the fraction of inventories lost due to storage costs. In the case of linear storage costs, s'_t is equal to a constant (call this s_1). Some researchers have modeled storage costs as being convex in the level of inventories. For example, convex inventory costs are the key factor driving Blinder and Fischer's (1981) model of the real business cycle. We capture these types of costs here by writing $s'_t = s_1 + (s_2/2) \cdot n_t$.[11]

For any cost minimizing firm that carries inventories between two periods, the marginal cost of producing an extra unit of output this period and holding it in inventories until next period must equal the expected marginal cost of producing an extra unit of output next period. The first order condition can be written as

$$MC_t = E_t\left[\frac{MC_{t+1}(1 - s_t)}{1 + \tilde{R}_t}\right] \qquad (2)$$

or

$$E_t\left[\frac{MC_{t+1}}{MC_t} \cdot \frac{(1 - s_t)}{1 + \tilde{R}_t}\right] = 1. \qquad (3)$$

Rational expectations implies

$$\frac{MC_{t+1}}{MC_t} \cdot \frac{(1 - s_t)}{1 + \tilde{R}_t} = 1 + \varepsilon_{t+1} \qquad (4)$$

where $E_t[\varepsilon_{t+1}] = 0$, i.e., ε_{t+1} is orthogonal to all information available at time t. The marginal storage cost, s_t, is equal to s_1 +

---

[11] If storage costs come in the form of depreciating inventories, then the accounting identity definition of output would be $y_t = x_t + (n_t - n_{t-1}(1-s'_{t-1}))$. In this paper, we construct output in the standard way: $y_t = x_t + (n_t - n_{t-1})$. If, rather than coming in the form of depreciated stocks, storage costs are actually paid out and these costs are proportional to the replacement cost of the goods, then our model and our constructed output measure are consistent with one another. In either of these cases the equations are correct when the IP measure of output is used. If the costs are paid out in dollars, in an amount related to the goods stored, our equation is approximately correct.

$s_2 \cdot n_t$.[12] The Euler equation (4) will not be satisfied if desired inventories are zero. We discuss this possibility below.

At this point it is worth pointing out the parallel between the production/storage problem of a cost minimizing firm and the consumption/saving problem of a utility maximizing consumer. The firm's problem is to minimize the expected discounted value of a convex cost function, subject to an expected pattern of sales and costs of holding inventories. The consumer's problem is to maximize the expected discounted value of a concave utility function, subject to an expected pattern of income and return to holding wealth. Not surprisingly, then, the solution to cost minimization yields a first-order condition analogous to the first-order condition implied by the stochastic version of the permanent income hypothesis (Hall (1978), Mankiw (1981), Hansen and Singleton (1983)), and we can apply the methods of that literature to testing the production smoothing model of inventories and output. Production, sales, inventories, the interest rate, and storage costs are analogous to consumption, income, wealth, the rate of time preference, and the return on wealth, respectively.[13] In the simplest version of this model, the real interest rate, the growth in the capital stock, and productivity growth are all constant over time. Given the production function that we employ, these assumptions imply that the expected growth in output is constant over time--i.e., real output follows a geometric random walk with drift. This is analogous to Hall's (1978) condition that consumption follow a random walk with drift.

To implement the model described above we need to specify the form of the cost function. We assume a standard Cobb-Douglas production function with m inputs (q_i, i = 1, ..., m). Let the last input (q_m) be

---

[12] If average storage costs s_t' are equal to $s_1 + (s_2/2) n_t$ this implies that marginal storage costs s_t are equal to $s_1 + s_2 n_t$.

[13] The non-negativity condition on inventories mentioned above is analogous to a borrowing constraint in the consumption literature. If time series/cross section data on firms were available, an approach similar to that of Zeldes (1985) could be applied here to test for the importance of this nonnegativity constraint on inventories.

the capital stock. In each period, the firm thus solves the following (constrained) problem:

$$\min_{\{q_1, q_2, \ldots, q_{m-1}\}} \sum_{i=1}^{m} w_i \cdot q_i \tag{5}$$

subject to

$$f(q_1, \ldots, q_m) = \mu \prod_{i=1}^{m} q_i^{a_i} = \bar{y},$$

$$q_m = \bar{q}_m,$$

where w_i and q_i are the price and quantity, respectively, of input i, and f is the production function. Note that the production function includes a productivity measure μ that may shift over time in deterministic and/or stochastic ways.[14] Define $A \equiv \sum_{i=1}^{m-1} a_i$. The one period (constrained) cost function from this problem is:

$$C(y) = w_m \cdot q_m + A \cdot q_m^{\frac{A-1}{A}} \left[\prod_{i=1}^{m-1} \left(\frac{w_i}{a_i}\right)^{a_i/A} \right] \cdot \mu^{-\frac{1}{A}} \cdot y^{\frac{1}{A}} \tag{6}$$

and the marginal cost function MC is:

$$MC(y) = q_m^{\frac{A-1}{A}} \left[\prod_{i=1}^{m-1} \left(\frac{w_i}{a_i}\right)^{a_i/A} \right] \cdot \mu^{-(1/A)} \cdot y^{(1-A)/A} . \tag{7}$$

Equation (7) can be used to calculate the ratio of marginal costs in t and t+1:

---

[14]Unlike some previous studies, we do not include costs of adjusting the level of output. As Maccini and Rossana (1984) point out, the costs of adjusting output presumably arise because of the cost of changing one or more factors of production. These costs may be important, but we do not attempt to model them here.

$$\ln(\frac{MC_{t+1}}{MC_t}) = [\sum_{i=1}^{m-1}(a_i/A)\cdot\ln(\frac{W_{it+1}}{w_{it}})] - (\frac{1-A}{A})\ln(\frac{q_{mt+1}}{q_{mt}}) + (\frac{1-A}{A})\ln(\frac{y_{t+1}}{y_t}) - \frac{1}{A}\ln\frac{\mu_{t+1}}{\mu_t}.$$
(8)

The next step is to derive an expression for the growth rate of output. We do so by taking logs of the Euler equation (4), taking a first order Taylor expansion of $\ln(1 - s_t)$ around $n_t = k$ for an arbitrary value of $k \geq 0$ and a second order Taylor expansion of $\ln(1 + \varepsilon_{t+1})$ around $\varepsilon_{t+1} = 0$, substituting in equation (8), and rearranging. This gives:

$$Gy_{t+1} = [(A/(1-A))] [-\ln(1-s_1-s_2k) - ks_2/(1-s_1-s_2k) - (1/2)\sigma_\varepsilon^2)]$$

$$+ (\frac{A}{1-A})[\ln(1 + \tilde{R}_t) - \sum_{i=1}^{m-1}(a_i/A)Gw_{it+1}] + Gq_{mt+1}$$

$$+ \frac{s_2}{1-s_1-s_2k}(\frac{A}{1-A})n_t + (\frac{1}{1-A})G\mu_{t+1}$$

$$+ (\frac{A}{1-A})[((1/2)\sigma_\varepsilon^2 - (1/2)\varepsilon_{t+1}^2) + \varepsilon_{t+1}] \quad (9)$$

where for any variable Z, $GZ_{t+1} = \ln(Z_{t+1}/Z_t)$. We have added and subtracted $(\frac{A}{1-A})(1/2)\sigma_\varepsilon^2$ from the equation, so that the last term in brackets in equation (9) has mean zero.

Discussion of the Model

Equation (9) is the basis of all the estimations performed in this paper. It says that the growth rate of output is a function of the real interest rate (where the inflation rate is used to calculate the real rate is a weighted average of the rates of inflation of factor prices), the growth in the capital stock, the level of inventories, productivity growth, and a surprise term. The key implication that we test in this paper is that no other information known at time t should help predict output growth.

As is well known, an advantage of estimating this Euler equation is that we avoid solving for firms' closed form decision rule for production. This allows us to step outside the linear-quadratic framework, and it allows us to estimate our model that includes stochastic input prices and interest rates. In addition, the Euler equation procedure yields testable implications for the growth rate rather than the level of output, so we do not need to assume that output is stationary around a deterministic trend. Our procedure is valid even if there is a unit root in the level of output, a condition that Nelson and Plosser (1982), find for example, characterizes aggregate output series.[15]

In setting up the model we have imposed the constraint that inventories are nonnegative, and we have indicated that the Euler equation is valid in a given period only if the nonnegativity constraint is not binding in that period. We should point out here a potential problem related to this nonnegativity constraint. Consider a certainty version of our model without the nonnegativity constraint imposed. Assuming that s_1 and s_2 are nonnegative, equation (3) implies that when inventories are positive firms want the level of marginal costs to rise over time. If the marginal cost _function_ is constant or falling (due to growth in the capital stock), this implies that output rises over time, i.e., that firms push production towards the future and run down inventory stocks today. In fact, only if inventories are negative could there be a steady state with constant marginal costs. This indicates that in a model in which the nonnegativity condition _is_ imposed, it will at times bind, and therefore the Euler equation will not be satisfied in some periods.[16]

To partially avoid this problem, we follow Blinder (1982) and allow s_1 to be negative. This captures the fact that at low but positive levels of inventories, increases in inventories may lower total costs, i.e., there may be a convenience yield to holding inventories. With s_1 sufficiently negative, there is a steady state in

---

[15] See Ghysels (1987) for an analysis of trends versus unit roots in manufacturing inventory and production data.

[16] Even under these assumptions, firms will in general choose to use inventories in some periods to smooth production, i.e., to build up positive inventories in the periods in which sales are especially high and run them down in periods in which sales are low.

the certainty version of the model that has a positive level of inventories.[17] Of course, this does still not imply, in the certainty or uncertainty version of the model, that inventories <u>never</u> hit the constraint.[18,19]

The model that we use allows for seasonal fluctuations in output growth in several ways. First, there may be seasonal movements in the observable or unobservable component of the productivity shifter. Second, there may be seasonals in the relevant input prices. Of course, it is not entirely accurate to describe these as determining the seasonal fluctuations in output growth, since in general equilibrium the seasonals in output growth and input prices are determined simultaneously. For an individual firm, however, and even for a 2-digit industry, the degree of simultaneity is likely to be small.

Rather than assuming that the productivity shifter μ is totally unobservable to the econometrician, we allow it to be a function of some observable seasonal variables and some unobservables. The observable variables are weather related: functions of current temperature and precipitation. It seems reasonable a priori that productivity would be affected by the current local weather. We write $\mu = e^{Z\gamma + \eta}$, where Z is a matrix of observable weather variables and η is the unobservable productivity shifter.

In the absence of shifts in the cost function (i.e., changes in μ), the model presented is a simple production smoothing model. For a given time path for the capital stock, the derived cost function is

---

[17] This can be seen by using (3), assuming no uncertainty, letting $R_t = R$, $n_t = n$, and setting $MC_{t+1} = MC_t$. Rearranging gives $n = (-R - s_1)/s_2$, which will be greater than zero if $s_1 < -R$.

[18] For a further discussion of this issue see Schutte (1983). Another factor in our model that tends to push inventories positive is sales growth, although this will be reversed to the extent that it is accompanied by growth in the capital stock.

[19] In the industries that we use to estimate the equation, industrywide inventories are always positive. This does not of course imply that inventories are always positive for every firm in these industries.

convex, inducing firms to try to spread production evenly over time.[20] When productivity is allowed to vary over time, the result is no longer a pure production smoothing model. Although our model is consistent with the variance of production exceeding the variance of sales, the convexity of the cost function remains and we continue to refer to the model as a type of production smoothing model.

Blinder (1986a) states that introducing (unobservable) cost shocks into the analysis makes his variance bounds inequality untestable, because one could explain an arbitrarily large variance of production relative to sales by assuming unobservable cost shocks with appropriately large variance. The approach that we adopt in this paper avoids this problem in two ways. First, we include <u>measurements</u> of a number of factors that might influence the marginal cost of production, and account for these in the analysis. Second, we show that under reasonable identifying assumption, the model described above has testable implications even in the presence of unobservable cost shocks. The most important assumption is that the unobserved component of productivity is uncorrelated with the component of sales that is predictable on the basis of information known at the time the firm makes its production decision. The testable implication is that once a number of cost variables are accounted for, the remaining movements in output should be uncorrelated with predictable movements in sales. In other words, even if production moves around a lot due to cost shocks, these movements should not be related to predictable movements in sales. This will be an especially useful test when applied to predictable <u>seasonal</u> movements in sales.

3. Identification and Testing

A. The General Approach

Equation (9), augmented to include the weather variables, can be written as:

$$Gy_{t+1} = [(A/(1-A)] \ [-\ln(1-s_1-s_2k) - (ks_2/(1-s_1-s_2k)) - (1/2)\sigma_\varepsilon^2]$$

---

[20] This smoothing that arises from a convex cost function is different than the smoothing induced by introducing costs of adjusting output (as in, for example, Eichenbaum (1984)). For further discussion, see Blanchard (1983).

$$+ \left(\frac{A}{1-A}\right)[\ln(1+\tilde{R}_t) - \sum_{i=1}^{m-1}(a_i/A)Gw_{it+1}] + Gq_{mt+1}$$

$$+ \frac{s_2}{1-s_1-s_2 k}\left(\frac{A}{1-A}\right)n_t + \left(\frac{1}{1-A}\right)(Z_{t+1}-Z_t)\cdot\gamma + \left(-\frac{1}{1-A}\right)G\eta_{t+1}$$

$$+ \left(\frac{A}{1-A}\right)[((1/2)\sigma_\varepsilon^2 - (1/2)\varepsilon_{t+1}^2) + \varepsilon_{t+1}] \qquad (10)$$

We cannot estimate this equation by OLS because the right-hand side variables are in general correlated with the expectations error. We therefore use an instrumental variables procedure to estimate the equation. To do so, we must choose instruments that are correlated with the included variables but not with the error term. Recall that the error term includes two components: the expectations error and the growth in the unobserved productivity shifter. Any variable that is known at time t will, by rational expectations, be orthogonal to ε_{t+1}.[21] However, rationality of expectations does not imply that $G\eta_{t+1}$ is orthogonal to time t information--it is possible that there are predictable movements in productivity growth. Note that a reasonable possibility is that the productivity measure follows a geometric random walk, in which case the growth in productivity is i.i.d. and therefore orthogonal to lagged information.[22]

Garber and King (1984) point out that a number of studies that estimate Euler equations assume that there are no shocks in the sector that they are estimating--effectively ignoring the identification issue. In this paper, we allow some measurable shocks to this sector, and we make the following identifying assumptions about the relationship between the unobserved cost shifter and the included instruments. (i) The unobserved productivity shifter (η) is

---

[21] We assume that production decisions for the month are made after information about demand and other economic variables is revealed, i.e., period t output decisions are made contingent on period t economic variables. An alternative assumption would be that production decisions are made before demand for the month is known. This creates a stockout motive for holding inventories (see Kahn (1986)). In Section 5, we also present results based on the alternative assumption that output must be chosen before demand for the period is known. For a further discussion of these timing issues, see Blinder (1986a).

[22] This assumption is made by Prescott (1986).

uncorrelated with lagged values of sales and with the part of current sales that was predictable based on lagged information. (ii) The growth of the productivity shifter is uncorrelated with lagged growth rates of input prices, lagged growth rates of output, and lagged interest rates.

We thus consider the following variables to be orthogonal to the error term in the regression: lagged growth in sales, lagged growth in output, lagged interest rate, lagged growth in factor prices, and lagged inventories. In some sets of results we relax the assumption that the lagged growth rate of output is uncorrelated with the growth in the productivity shifter. To test the model, we first estimate equation (10) with instrumental variables, including as instruments the variables in the above list. Since there are more instruments than right-hand side variables, the equation is overidentified. We then test the overidentifying restrictions by regressing the estimated residuals on all of the included instruments (including the predetermined right-hand side variables). The quantity T times the R^2 from this regression is distributed χ^2_j, where T is the number of observations and j is the number of overidentifying restrictions. One possible alternative hypothesis to our null is that firms simply set current output in line with current sales. In this case, we would expect the lagged growth rate of sales to enter significantly in our test of the overidentifying restrictions.

B. **Seasonality and Identification**

It is possible that there are seasonal movements in productivity that are not captured by the weather variables. One possible way to capture these would be to allow the productivity measure to be an arbitrary function of seasonal dummies. We do this in our first set of results by including seasonal dummies in the estimation of equation (10).[23] This gives the same results as first regressing all of the

---

[23] The fact that we include seasonal dummies does not mean that we assume purely deterministic seasonality. Since the right hand-side variables may exhibit stochastic seasonality, our model allows for both stochastic and deterministic seasonality in output growth. We should also note that because we are working in log first differences, using additive seasonal dummies allows for multiplicative seasonality in

variables on seasonal dummies, and then using the residuals from these regressions for estimation purposes.

In order to examine whether the use of X-11 adjusted data has been responsible for the rejections by others of the production smoothing model, we compare the tests of the model using seasonally unadjusted data and seasonal dummies to the tests using X-11 seasonally adjusted data.

When we include seasonal dummies in equation (10), we lose all power to test the model at seasonal frequencies, i.e., we cannot test whether the seasonal movements in the data are consistent with the model. In the latter part of the paper, therefore, we make the stronger identifying assumption that seasonal shifts in productivity not captured by weather variables are uncorrelated with the instruments used to estimate equation (10).[24] Under this assumption, we can exclude seasonal dummies and perform two further tests that directly use seasonal fluctuations in the data. We test the implication that once the other factors in the cost function are taken into account, the remaining movements in output should be uncorrelated with the seasonal movements in sales. This is a strong implication of the production smoothing model that has not been tested to date. In addition, we examine whether the model fits at purely seasonal frequencies. We describe these latter two tests in Section 6.

4. **The Data**

This section describes the data set that we employ. There are a number of technical issues to be considered with respect to both the adjusted and unadjusted data on inventories and production; we discuss these in detail below. Readers who are not interested in these details can skip to Section 5.

The equations are estimated using monthly data from May 1967

---

output.

[24] In the section below on seasonal results, we discuss the circumstances under which this assumption might not hold.

through December 1982.[25,26] Data on inventories and shipments at the 2-digit SIC level for 20 industries were obtained from the Department of Commerce. We estimate the equations only on the six industries identified by Belsley (1969) as being production to stock industries. The inventory data are end of month inventories of finished goods, adjusted by the Bureau of Economic Analysis (BEA) from the book value reported by firms into constant dollars.[27,28] We follow West (1983) and adjust the BEA series from "cost" to "market," so that shipments and inventories are in comparable units. Shipments data are total monthly figures in constant dollars.

Two different measures of production are used. The first comes from the identity that production of finished goods equals sales plus the change in inventories of finished goods. Commerce Department data for sales and the change in inventories are used to compute this production measure (which we call "Y4"). The second measure of production used is the Federal Reserve Board's index of industrial production (IP), also available at the 2-digit SIC level.

In principle, the two production series measure the same variable and should therefore behave similarly over time. As documented in Miron and Zeldes (1987), however, the two series are in fact quite different. For the six industries studied here, the correlations

---

[25] Most of our data run through December 1984, but we only have weather data through December 1982.

[26] A month seems like a reasonable planning horizon for a firm, but there is no obvious reason why it need be so. For a discussion of time aggregation issues in inventory models, see Christiano and Eichenbaum (1986).

[27] This adjustment attempts to take into account whether firms used LIFO or FIFO accounting. See Hinrichs and Eckman (1981) for a description of how the constant dollar inventory series are constructed. See Reagan and Sheehan (1985) for a presentation of the stylized facts of these series at an aggregate (durables and non-durables) level.

[28] There is some disagreement over whether it is appropriate to use finished goods inventories only (West (1986)) or finished goods plus work in progres inventories (Blinder (1986a)). We estimate the equations separately for each definition. See footnote 32 in Section 5.

between growth rates of the two series range from .8 to .4 for the seasonally unadjusted data, and from .4 to less than .1 for the seasonally adjusted data. The serial correlation properties and seasonal movements of the two series are also different. Since we have not resolved this discrepancy, we present results based on both output measures.

The nominal interest rate is the yield to maturity on Treasury Bills with one month to maturity as reported on the CRSP tapes. The marginal corporate tax rate series is the one calculated by Feldstein and Summers (1979). The input price series are wages, the price of crude materials for further processing, and energy prices, representing the three largest variable inputs in the production process. Wages (average hourly earnings) and industrial production at the 2-digit SIC level, and aggregate measures of energy prices (the PPI for petroleum and coal products) and raw materials prices are available from the Citibank Economic Database.

The capital stock enters our equations as the number of machine days used per month. Since we did not have access to industry capital stock data, we model the growth in the capital stock as a constant plus a function of the growth in the number of non-holiday weekdays in the month. Any remaining month to month variation in the growth in the capital stock is included in the error term.

The weather data include estimates of total monthly precipitation and average monthly temperature. We construct a different temperature and precipitation measure for each industry, equal to weighted averages of the corresponding measures in the different states. The weights are equal to the historical share of the total shipments of the industry that originated in each state.[29] To capture nonlinearities, we also include the weighted average of squared temperature, squared precipitation, and the cross-product of temperature and precipitation. Given our functional form assumptions, the first differences of these variables enter equation (10).

Seasonality

---

[29] The weights change every five years but always correspond to averages of previous (never future) years.

Whenever possible, we obtained both seasonally adjusted (SA) and seasonally unadjusted (NSA) data. The BEA reports real shipments and inventories data, but these constant dollar series are only available on a SA basis.[30] The Bureau of the Census reports NSA and SA current dollar shipments series and book value inventories series. As in Reagan and Sheehan (1985) and West (1986), we estimate the real NSA inventory series by multiplying the real SA series by the ratio of book value NSA to book value SA, thus putting back in an estimate of the seasonal. (Another way of thinking of this is that we deflate the book value NSA series by the ratio of the book value SA to real SA series.) We estimate real NSA shipments by multiplying the real SA series by the ratio of nominal NSA shipments to nominal SA shipments. These procedures assume that there are no seasonal movements in the factors that convert from book value to current dollar value or in the deflators used to convert the series from current dollar to constant dollar. An additional adjustment that we considered was to multiply the above series by the ratio of the SA to NSA PPI series for the finished goods, in order to adjust for the seasonal in the deflators. We found statistically significant evidence of seasonality in the price indexes in three out of six industries. However, the magnitudes of the seasonal movements in these prices are much smaller than in the corresponding quantities. We estimated the specifications in Tables 2 and 5 both with and without this adjustment and the results were virtually identical to each other.[31] We report only the results without this last adjustment.

The IP data are available both NSA and SA, and the energy price series, wage rates, raw materials prices and interest rates are all unadjusted.

5. Basic Results

---

[30] The reason for this has to do with the technique used to construct the constant dollar figures. The disaggregated nominal series are first seasonally adjusted, then deflated and then aggregated.

[31] We estimated the equation over a shorter sample period for the food, chemicals, and petroleum industries because seasonally adjusted PPIs were unavailable for part of the sample period.

In this section, we examine the basic results from estimating equation (10) and testing the implied overidentifying restrictions. In order to determine whether the use of X-11 adjusted data has been responsible for previous rejections of the production smoothing model, we run the same set of tests with (i) the standard X-11 seasonally adjusted data and (ii) seasonally unadjusted data plus seasonal dummies.

A summary of the results is presented in Table II.[32] There are four sets of results, since we carry out the estimation with both unadjusted and adjusted data, and we do this for both the Y4 and IP measures of output. In the first line of each set of results, we list the variables that entered equation (10) at a significance level of 5%. In the second line of each set, we present the R^2 from the regression of the residuals on all the instruments. Recall that $T \cdot R^2$ is distributed χ_j^2, where j is the number of overidentifying restrictions. On the same line, we report the marginal significance level of the test statistic $T \cdot R^2$. In the last line of each set, we list the variables that entered this auxiliary test significantly.

We make the following observations about the results. First, in no case does the interest rate or the growth rate in energy prices enter equation (10) significantly. In about one third of the cases, the growth in raw materials prices enters significantly, but usually with the wrong sign. Wage growth enters significantly only four times, twice with the wrong sign. Thus, the signs and statistical significance of the coefficient estimates are not supportive of the model.

The second observation we make is that the data reject the overidentifying restrictions on the model in all cases using the Y4 data, and in two-thirds using the IP data. For the Y4 data, the rejections are about as strong using seasonally adjusted and seasonally unadjusted data. For the IP data, the rejections are not quite as strong overall using the seasonally unadjusted data. On the whole,

---

[32] Most of these estimations were also done using the sum of finished goods inventories and work in progress inventories as the definition of inventories. The results were almost identical to those reported in the text.

TABLE II

REGRESSION RESULTS, EQUATION (10) SEASONAL DUMMIES IN EQUATION AND INSTRUMENT LIST

		Food	Tobacco	Apparel	Chemicals	Petroleum	Rubber
Y4, NSA	What enters (10) significantly?	sd,rm	tem,-tem2	sd	-sd,tem -tem2,-pre	-w,-day -tem2,-pre2	sd
	R^2, Significance level	.17,.000	.23,.000	.09,.050	.15,.001	.13,.004	.16,.000
	What is significant in test of OIR's?	we-1,'-we-2	tem-1	---	-y-1,'x-1 -n-2,'n-3,sd	we-2	w,sd
Y4, SA	What enters (10) significantly?	---	rm	-n-1, rm	rm	-w,-day -pre2,tpr	w
	R^2, Significance Level	.10,.027	.14,.002	.11,.014	.14,.002	.14,.002	.11,.014
	What is significant in tests of OIR's?	---	---	---	---	---	---
IP, NSA	What enters (10) significantly?	sd	-sd,w,-pre2	sd,day-1	sd, rm	sd,-pre2	rm
	R^2, Significance Level	.16,.000	.16,.000	.09,.050	.04,.583	.08,.090	.02,.926
	What is significant in test of OIR's?	-y-1,'-w-2	-y-1,'sd	-y-1	---	---	---
IP, SA	What enters (10) significantly?	---	-pre2	-pre2,day-1	tem,-tem 2	-day-1	rm, tem,tem2
	R^2, Significance Level	.15,.001	.28,.000	.09,.050	.13,.004	.09,.050	.06,.257
	What is significant in test of OIR's?	-y-1,-we -tem-1,tpr-1	-y-1,-y-2	-y-1	x-1,x-2	---	-n-1

Notes:
1. The sample period is 1967:5-1982:12.
2. The first line of each set of results lists the variables that entered equation (10) at the 5% significance level. We list seasonal dummies if one or more of the eleven dummies entered significantly.
3. The second line gives the R^2 from the regression of the residuals on the instruments, as well as the marginal significance level of this statistic. The quantity $T \times R^2$ is distributed χ_j^2, where j is the number of overidentifying restrictions and T is number of observations. In the results presented here, there are 9 such restrictions.
4. The third line lists the variables that entered the regression of the residuals on the instruments at the 5% significance level.
5. w = wage growth, sd = seasonal dummies, y = output growth, x = sales growth, day = number of production days, we = energy p growth, rm = raw materials price growth, n = inventories, r = interest rate, pre = change in precipitation, pre2 = change in precipitation squared, tem = change in temperature, tem2 = change in temperature squared, tpr = change in temperature precipitation.
6. A (-) before variable indicates that the sign of the coefficient was negative.
7. A subscript of -1 on a variable means that it is dated 1 periods earlier than the dependent variable.

there is little evidence that the use of unadjusted data with seasonal dummies provides better results than using seasonally adjusted data.

Finally, note that in approximately half of the cases, at least one of the five weather variables enters the equation significantly. Even after including seasonal dummies, the weather has a significant influence on production in certain industries (tobacco, chemicals, and petroleum).

Thus far, we arrive at a negative assessment of the model for two reasons. First, the overidentifying restrictions are typically rejected. Second, the signs of the coefficient estimates are not sensible and rarely significant. Proponents of the model might make the following argument against these two reasons, respectively. First, the instrument list may include variables that are correlated with the error term even under the null hypothesis, thus invalidating the tests of the overidentifying restrictions. Second, the instruments may not do a very good job of explaining the right hand side variables. If this is the case, one should not expect the parameter estimates to be statistically significant, even under the null. We discuss each of these arguments in turn.

There are two circumstances in which the instrument list employed, consisting of lagged values of production, sales, input prices, and inventories, may be correlated with the error term. First, lagged output growth may not be a valid instrument, even if other lagged variables are, because productivity growth might be serially correlated. Since productivity growth is correlated with output growth, this implies that lagged output growth will also be correlated with contemporaneous productivity growth (a component of the error term), making it an invalid instrument.

Second, if firms do not have complete current period information when they make their output decisions for period t, then variables dated time t may not be valid instruments. This could arise because firms do not know the demand for their own products for the period before choosing output (as in Kahn (1986) or Christiano (1986)). Alternatively, firms may know the total demand for their product, but not the aggregate component of demand. Since we are using data on firms aggregated to the industry level, this too might invalidate the use of time t instruments (see Goodfriend (1986)).

In order to take account of these possibilities, we have estimated equation (10) using two alternative instrument lists. The first excludes production from the instrument list and includes extra lags of sales. The second list excludes all variables dated time t and includes extra lags of the variables at earlier dates.

When we employ the first alternative instrument list we reject the overidentifying restrictions significantly less often than with the list used in our basic results. In this case, the restrictions are rejected in a majority of cases for the Y4 data, but never for the IP data. When we employ the second alternative instrument list, we never reject the overidentifying restrictions. In both cases, however, we almost never find that the input price variables, the interest rate, or the level of inventories enter statistically significantly with the correct sign.

This brings us to the second issue. It is possible that we are not finding that expected changes in input prices affect the timing of production because there are no expected changes in input prices. That is, the instruments that we employ, either in our basic results or alternative results, may be of such poor quality that they have no explanatory power for the right-hand side variables in equation (10). If this is the case, the failure of these input prices to explain the pattern of production is not evidence against our model.

It is easy to check this possibility by examining directly the explanatory power of the instruments. For all three instrument lists, we find the following: there is statistically significant explanatory power in the instruments about half the time for wages; all the time for interest rates, energy prices, and all five weather variables; and almost never for raw materials prices. Thus, with the exception of raw materials prices, the failure of input prices to explain production in any of our results is valid evidence against the model.

To summarize, with our basic instrument list the results provide evidence against the production smoothing model, even when it is expanded to incorporate a stochastic interest rate, measurable and unmeasurable cost shocks, and non-quadratic technology. When two weaker sets of identifying restrictions are used, there is substantially less statistical evidence against the model, but there is still no evidence that it describes an important aspect of firm behavior. Using seasonally unadjusted data and seasonal dummies does little better than using X-11 adjusted data.

6. Seasonal-Specific Results

In this section we examine the extent to which the seasonal fluctuations in production, shipments and inventories are consistent with the production smoothing model. The results presented above incorporate seasonal fluctuations into the analysis by using seasonally unadjusted data and including seasonal dummies and weather variables in the equations. This approach does not tell us to what extent the seasonal movements in interest rates or input prices determine the seasonal movements in output growth, nor does it answer the question of whether the seasonal movements in the data themselves satisfy the production smoothing model. In order to answer these questions, we cannot include seasonal dummies in equation (10), and must therefore assume that any fluctuations (seasonal and nonseasonal) in the productivity measure not captured by the weather variables are orthogonal to the instruments used.

Before describing our formal tests, it is useful to consider a set of stylized facts about the seasonality in production, inventories, and sales. We saw in Table I that the seasonal variation in the data is large relative to the non-seasonal variation. In Table III, we present estimates of the ratio of the variance of production to the variance of sales, and we include estimates based separately on the seasonal and non-seasonal variation. Following Blinder (1986a), these numbers are

TABLE III

VARIANCE OF PRODUCTION DIVIDED BY VARIANCE OF SALES

	Food	Tobacco	Apparel	Chemicals	Petroleum	Rubber
Nonseasonal	1.22	1.84	1.32	1.01	0.91	1.13
Seasonal	1.71	4.71	0.58	0.72	2.73	0.99
Total	1.50	2.53	0.80	0.89	0.94	1.09
Nonseasonal	0.48	0.58	1.20	0.83	0.48	0.95
Seasonal	1.64	6.28	0.21	0.18	7.91	0.61
Total	1.14	1.95	0.50	0.55	0.59	0.86

Notes:
1. The sample period is 1967:5-1982:12.
2. The estimation procedure is the following. For both shipments and production, the log level is regressed on a constant, time and time trend that is one beginning in October, 1973. The coefficients are estimated by GLS, assuming a second order autoregressive process for the error term. The antilogs of the fitted values of this regression are then subtracted from the actual data, in levels, to define the detrended data. The seasonal and nonseasonal variances are calculated as the variance of the fitted values and residuals, respectively, of a regression of the detrended series on seasonal dummies.
3. We convert the IP measure from an index to a constant dollar figure by multiplying it by the ratio of average Y4 to average IP.

based on detrended levels rather than growth rates.[33,34] As we have discussed above, if cost shocks are assumed to be "small," the production smoothing model restricts these ratios to be less than one. We focus here on the ratio of the seasonal variances. For three of the six industries, we estimate this ratio to be greater than one.[35]

While one could interpret a ratio greater than one as a rejection of the production smoothing model, there is no reason to expect the above ratio to be less than one if there are seasonal shifts in the cost function. Even in this case, however, there is information to be learned from examining the seasonal movements. Whether or not seasonal shifts in productivity affect the seasonal pattern of production, there is no reason to expect that seasonal pattern to match the seasonal pattern of sales. Figures 1-6 show the seasonal patterns in output and shipments for the six industries we examine and document behavior potentially problematic for the production smoothing model.[36] <u>The seasonal movements in output and sales are in fact very similar</u>. The implication of these graphs is that inventories do not appear to be playing the role of smoothing seasonal fluctuations in sales.

---

[33] Along the lines of Blinder, we use the following procedure to obtain detrended levels of the data. The log of each series is regressed on a constant, time, and a dummy variable that is one beginning in October 1973. The coefficients are estimated by GLS, assuming a second order autoregressive process for the error term. The antilogs of the fitted values of this regression are then subtracted from the levels of the raw data to define the trended data. We convert the IP measure from an index into a constant dollar figure by multiplying it by the ratio of average Y4 to average IP (i.e., we set the average of the two series equal to each other). We apply the detrending procedure to the resulting IP, as well as Y4 and shipments. We then regress the detrended series on a constant and eleven seasonal dummies. The seasonal and non-seasonal variances are estimated using the fitted and residual values of this regression, respectively.

[34] In the last section of his paper, West (1986) describes a variance bounds test that includes deterministic seasonal variations in the data. He found that the variance bounds were rejected for each of the three industries that he examined.

[35] We examine these ratios for seasonally adjusted data in Miron and Zeldes (1987), and find significant differences between the ratios based on IP and Y4 data.

[36] The seasonal coefficient plotted for each month is the average percentage difference in that month from a logarithmic time trend.

In the tests we present in this section, we formalize this observation. First, we test whether the contemporaneous seasonal movement in sales growth helps predict residual output growth, once the movements in factor prices, the weather, and lagged inventories are taken into account. To do this, we use the same procedure as in Section 5, except that seasonal dummies are excluded from the regression and the instrument list, and the seasonal component of contemporaneous sales growth is added to the instrument list. It is unusual when running this type of orthogonality test to include as an instrument a contemporaneous variable, but since this series is deterministic, it is part of the lagged information set. Since it is also assumed orthogonal to the unobservable productivity shifter, it is a valid instrument.[37]

The interpretation of this procedure is the following. By excluding seasonal dummies from the equation, we force the seasonal and nonseasonal movements in the right-hand side variables to effect output growth via the same coefficients. Given this restriction, we are then testing whether the part of output growth not explained by these variables is correlated with the seasonal component of sales growth. This allows us to compare the seasonals in sales and output, after taking into account the measured seasonality in factor costs, the weather, and the level of inventories.

This test of the production smoothing model using the seasonal fluctuations does involve one important maintained hypothesis, namely that the coefficients on the seasonal and non-seasonal components of input prices and the weather are the same. In our last set of tests, we relax this assumption and test whether the seasonal movements in the data, taken by themselves, are consistent with the model. This is accomplished as follows. The first step is to construct the seasonal component of each of the relevant variables (output growth, input prices, weather variables, etc.) by regressing them on seasonal dummies and calculating the fitted values of these regressions. We then regress the seasonal component of output growth on the seasonal in input prices, weather, the level of inventories, and the

---

[37] The series we actually use is, of course, the estimated rather than the true seasonal in sales growth.

contemporaneous seasonal component of sales, and we test the restrictions implied by the model that this last coefficient should be zero.[38,39]

The results are summarized in Tables IV and V. Table IV presents the same type of information as Table II, but it includes the t-statistic on the seasonal component of contemporaneous sales in the test of the overidentifying restrictions. Table V is also set up similarly to Table II, but it simply reports whether the seasonal in sales significantly affects the seasonal in output growth, after controlling for the seasonal movements in input prices, the weather, and the level of inventories.

In both tables, there is striking evidence against the production smoothing model. In Table IV, we reject the overidentifying restrictions in every instance. In most cases, the seasonal component of sales is significantly correlated with the movement in output, even after taking account of any seasonals in input prices, lagged inventories, and the weather. This is true for five out of six industries using at least one of the output measures, and for three of the six industries using both output measures.

---

[38] Since we found there to be essentially no seasonality in energy prices, raw materials prices, or interest rates, we exclude these variables.

[39] We implement the procedure above by estimating equation (10), with sales growth included, using seasonal dummies as the only instruments. The coefficient estimates are numerically identical to those produced by the procedure described in the text, but this instrumental variables procedure produces correct standard errors. The resulting t-statistic on sales growth is reported in Table 5.

FIGURE 1

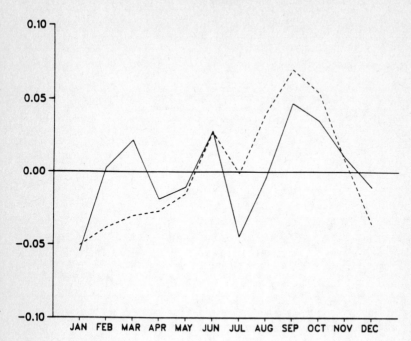

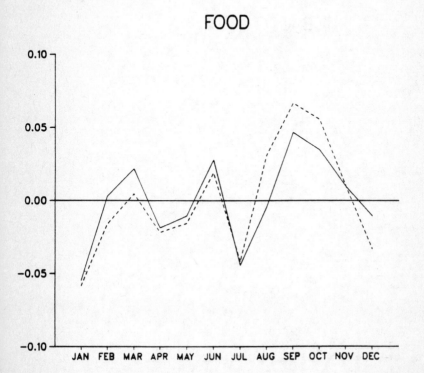

FIGURE 2

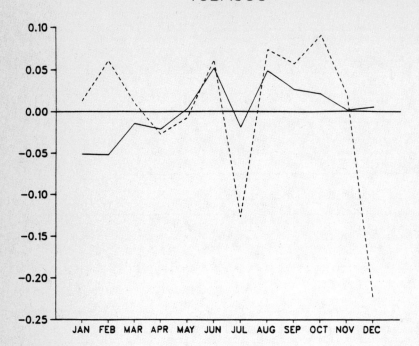

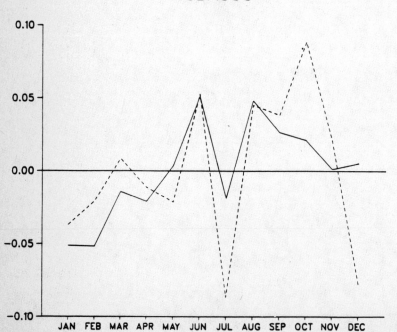

FIGURE 3

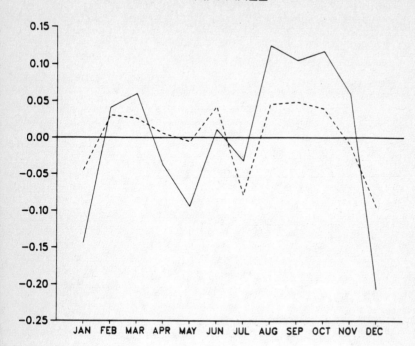

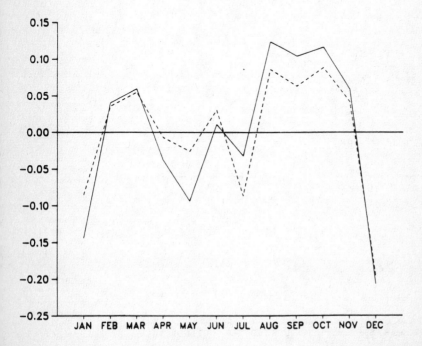

FIGURE 4

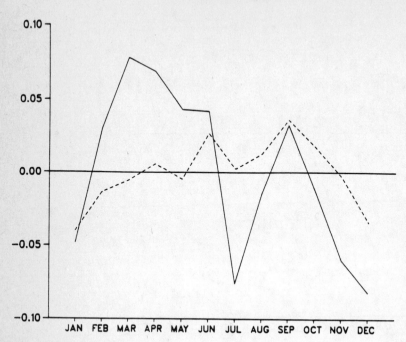

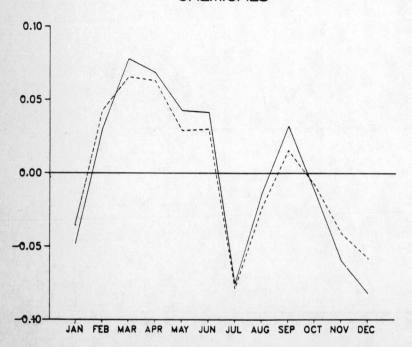

FIGURE 5

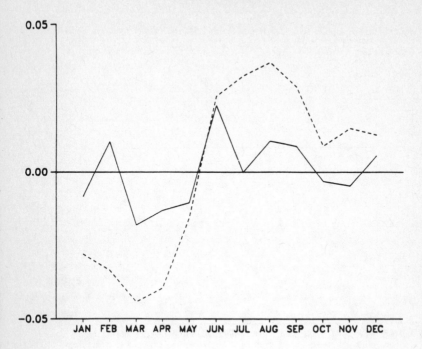

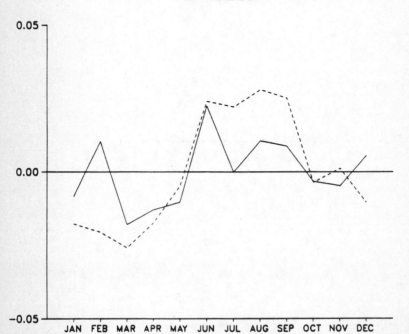

FIGURE 6

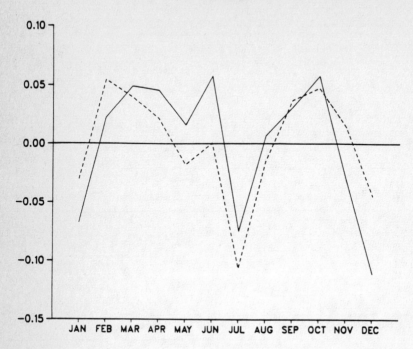

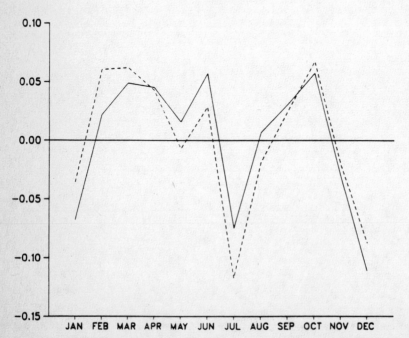

When we redo the estimates in Table IV using the alternative instrument lists discussed above (leaving out time t variables or lagged output growth) we again reject the overidentifying restrictions and find that the seasonal in sales growth is significantly correlated with the residual output growth in most cases.

In Table V (the seasonal-only results) the seasonal in sales growth is statistically significant in five out of six cases for the Y4 measure of output, and in four out of six cases for the IP measure. Variables other than sales almost never enter significantly.

These results on the behavior of production and sales at seasonal frequencies are perhaps the most problematic yet presented for the production smoothing model. To a large extent, firms appear to be choosing their seasonal production patterns to match their seasonal sales patterns, rather than using inventories to smooth production over the year. Moreover, since the seasonal variation in production and sales growth generally accounts for more than 50% of the total variation in these variables, this problematic behavior is a quantitatively important feature of the data.

A key assumption that we have made here is that the seasonal in the productivity shifter is uncorrelated with the seasonal in demand. Are there circumstances under which this assumption would not hold? An example that comes to mind is the case of an economy-wide seasonal in labor supply, namely that individuals, all else equal, would rather take vacations in certain months. This would induce a corresponding seasonal in output. If each industry's output is an input into another industry, then we might expect to see a corresponding seasonal in shipments, leading optimally to the same seasonal patterns in output and shipments.

Theoretically, our approach accounts for this by including the wage as a determinant of desired production. However, if the measured wage differs from the true shadow cost of utilizing labor, then the residual will include the seasonal in labor supply and therefore still be correlated with the seasonal in shipments. This explanation suggests that we should see the same seasonal movements in output in all industries. In Figures 1-6, we do see common seasonal patterns in output across industries, but we also see a fair amount of seasonal

TABLE IV

REGRESSION RESULTS, EQUATION (10) SEASONAL DUMMIES EXCLUDED,
SEASONAL COMPONENT OF SALES IN INSTRUMENT LIST

Y4,NSA	What enters (10) significantly?	-w,tem,-tem2	-w	w,tem -tem2	-w,tem -pre,-tem2	-w,pre2 -day	-w,tem tem2,-pre
	R^2,Significance Level	.52,.000	.16,.001	.14,.003	.31,.000	.18,.000	.26,.000
	What is significant in test of OIR's	tem-1,-tem2 -x-2,-rm-1 xseas	tem-1	xseas	tem-1,xseas	we-2	xseas
	t-stat on xseas	3.51	0.51	2.84	4.14	1.87	3.41
IP,NSA	What enters (10) significantly?	-w,tem,-tem2	-w	w,rm tem,-tem2	tem,-tem2 -pre	-w,pre2	tem,-tem2
	R^2,Significance Level	.30,.000	.14,.003	.12,.012	.15,.002	.13,.007	.24,.000
	What is significant in tests of OIR's?	-Y-1,-Y-2, -W-2,xseas	-Y-1,-W-2	-Y-1,xseas	----	-Y-1,-X-1 xseas-1	xseas
	t-stat on xseas	2.82	0.99	2.62	1.82	3.39	3.41

Notes:

1. The sample period is 1967:5-1982:12.
2. The first line of each set of results lists the variables that entered equation (10) at the 5% significance level.
3. The second line gives the R^2 from the regression of the residuals on the instruments, as well as the marginal significance level of this statistic. The quantity T x R^2 is distributed v_j, where j is the number of overidentifying restrictions and T is number of observations. In the results presented here, there are 10 such restrictions.
4. The third line lists the variable that entered the regression of the residuals on the instruments at the 5% significance le
5. w = wage growth, y = output growth, x = sales growth, day = number of production days, we = energy price growth, rm = raw materials price growth, n = inventories, r = interest rate, pre = change in precipitation, pre2 = change in precipitation, pre change in precipitation squared, tem = change in temperature, tem2 = change in temperature squared, tpr = change in temperatur percipitation, xseas = seasonal component of sales growth (defined as the fitted values from a regression on seasonal dummies)
6. A (-) before variable indicates that the sign of the coefficient was negative.
7. A subscript of -1 on a variable means that it is dated -1 periods earlier than the dependent variable.

TABLE V

REGRESSION RESULTS, EQUATION (10) SEASONAL COMPONENTS ONLY

	Food	Tobacco	Apparel	Chemicals	Petroleum	Rubber
Y4 What enters (10) significantly?	-n,x	x	x	x	x	x
t stat on x	9.21	4.91	1.98	11.69	1.10	19.99
IP What enters (10) significantly?	x	-pre,x	----	x	----	x
t stat on x	3.19	3.39	1.14	2.31	.62	3.97

Notes:
1. The sample period is 1967:5-1982:12.
2. The coefficients were obtained by estimating equation (10), with sales growth included, using seasonal dummies only as instruments. This gives coefficient estimates numerically identical to the procedure described in the text, but it produces correct standard errors. The t statistic on sales growth reported in the table is from this instrumental variables regression.
3. The first line of each set of results lists the variables that entered equation (10) at the 5% significance level.
4. w = wage growth, y = growth output, x = sales growth, day = change in number of production days, n = inventories, pre = cha in precipitation, pre2 = change in precipitation squared, tem = change in temperature, tem2 = change in temperature squared, t change in temperature* precipitation.
5. A (-) before variable indicates that the sign of the coefficient was negative.

movement that is different across industries.

It is not clear what conclusion to draw from this discussion. It is possible that the hypothesis proposed above is the explanation for the seasonal results. If so, we should ask whether the same type of arguments could be made about non-seasonal movements, i.e., whether we believe that the failure of the production smoothing model at non-seasonal frequencies is due to economy-wide changes in desired labor supply that are not captured by measured wages.

7. Conclusions

The results presented above show a strong rejection of the production smoothing model. This is despite the fact that we have extended the standard model considerably, by allowing for non-quadratic technology, a stochastic interest rate, convex costs of holding inventories, and measurable and non-measurable cost shocks, and by including seasonal fluctuations explicitly. Although previous work has examined many of these features, non has simultaneously allowed for all of them.

The rejections of the basic production smoothing model that we report are robust with respect to the treatment of seasonal fluctuations. To begin with, we reject the model about as strongly when we treat seasonality in the standard way, by using adjusted data, as when we treat it more explicitly by specifying the economic sources of the seasonal movements in production and inventories. Even more surprisingly, our results show that the seasonal movements in production, inventories and shipments are inconsistent with the basic model. Specifically, the seasonal component of output growth, even after adjusting for the seasonality in interest rates, wages, energy prices, raw materials prices, and the weather, is still highly correlated with the seasonal component of sales growth, contrary to the prediction of the model.

We conclude the paper by discussing what we believe to be the implications of our results for a number of hypotheses that have been offered for the failure of the production smoothing model. We first discuss those hypotheses on which our results provide direct evidence and then turn to more indirect implications.

Our results provide direct evidence that the limited role given to cost shocks in previous papers is not the major reason for the rejections of the model. In this paper we have included a more general set of cost shocks than in earlier work, and we still find that the data reject production smoothing. Moreover, we find relatively little evidence that cost shocks play any role in determining the optimal timing of production. It is possible, of course, that we have omitted the "key" cost shock, or that one of our identifying assumptions is invalid. We believe, however, that the set of costs we have included covers all of the major ones, and we think that the identifying assumptions we make are minimally restrictive. It seems to us unlikely, therefore, that the treatment of cost shocks is a major factor in explaining the poor performance of the model.

The second area in which our results provide direct evidence is on whether the inappropriate use of data seasonally adjusted by X-11 has been responsible for the failure of the model. As we discussed above, X-11 data are (approximately) a two-sided moving average of the underlying seasonally unadjusted data. This means that such data likely violate the crucial orthogonality conditions that are tested in the kinds of models considered above, even if the unadjusted data satisfy them. Although it seemed likely to us on a priori grounds that the use of X-11 adjusted data was a major problem, our results indicate otherwise. The particular method of treating seasonal fluctuations does not appear crucial to an evaluation of the model.

So much for direct implications. We now turn to more indirect implications, specifically, the implications of our tests using the seasonal movements in the data. These implications are subject to the critique that the production smoothing model may fit differently at different frequencies, in which case we may not be able to learn about the validity of the model at nonseasonal frequencies from its performance at seasonal frequencies. However, to the extent that the same model is underlying the different movements in the data, we can draw the following conclusions.

To begin with, since seasonal fluctuations are anticipated, it seems unlikely that the failure of the model at seasonal frequencies could be due to any kind of irrationality or disequilibrium. If so, this rules out a large class of possible explanations of the failure of the model.

A second issue that is illuminated by our seasonal specific results is that of costs of changing production. We have omitted costs of changing production (or, more generally, costs of changing inputs) from our specification above; the addition of these costs might "help the data fit the model." We regard this tactic as unsatisfactory, however. The fact that there are extremely large seasonal changes in the rate of production makes it seem quite unlikely that there are large costs of adjustment, although it is true that costs may be lower when they are anticipated.

Blinder (1986a) suggests that the production smoothing model could be saved by including persistent demand shocks and small cost shocks. Even if the non-seasonal movements in sales are very persistent, however, the same is not true of the seasonal movements. Therefore, our seasonal results suggest that Blinder's explanations will not suffice to "save" the production smoothing model.

Finally, our seasonal specific results allow us to rule out a concern regarding the choice of appropriate instruments. In our estimation, we assume that firms know current demand, and therefore time t sales is a valid instrument. In contrast, others such as Kahn (1986), assume that firms do not know the level of current period demand when they choose the current period level of production. If this assumption is a more appropriate abstraction, then our general results are inconsistent. When we correct for this by using only variables dated $t-1$ and earlier as instruments, we can no longer reject the model. However, it is still valid to include the seasonal components of contemporaneous sales growth, since the seasonal component of demand would be known even if the overall level were not. Since the results from this test show a strong rejection of the model, this suggests that the assumption that firms observe demand before choosing output is not, by itself, to blame.

What remains, then, as a possible explanation for the failure of the production smoothing model? There are two main possibilities: non-convexity of the cost function, and stockout costs. Giving up convexity of costs is unappealing because it requires also giving up much of neo-classical theory. This does not mean it is not the correct explanation; it simply suggests that we should turn to it only as a last resort. We end, therefore, by discussing the role of stockout costs.

An important maintained assumption above is that firms always hold positive inventories, which implies that firms do not stock out. Total inventories for each industry are always positive in our data, but this may not be the case for each individual firm or product. Kahn (1986) presents a model in which, because stockouts are costly, firms may not smooth production.[40] However, there is as yet relatively little direct evidence that stockout costs are high, or that firms cannot simply hold unfilled orders as a type of negative inventories. This line of research deserves further attention, in particular direct empirical testing.

---

[40]Abel (1985) also examines production smoothing in a model with stockouts.

References

Abel, Andrew B. (1985): "Inventories, Stock-Outs and Production Smoothing," Review of Economic Studies, 52, 283-293.

Belsley, David A. (1969): Industrial Production Behavior: The Order-Stock Distinction, Amsterdam: North-Holland.

Blanchard, Olivier J. (1983): "The Production and Inventory Behavior of the American Automobile Industry," Journal of Political Economy, 91, 365-400.

Blinder, Alan S. (1986a): "Can the Production Smoothing Model of Inventory be Saved?," Quarterly Journal of Economics, 101, 431-454.

Blinder, Alan S. (1986b): "More of the Speed of Adjustment in Inventory Models," Journal of Money, Credit, and Banking, 18, 355-365.

Blinder, Alan S. (1982): "Inventories and Sticky Prices: More on the Microfoundations of Macroeconomics," American Economic Review, 72, 334-348.

Blinder, Alan S. and Stanley Fischer (1981): "Inventories, Rational Expectations, and the Business Cycle," Journal of Monetary Economics, 8, 277-304.

Christiano, Lawrence J. (1986): "Why Does Inventory Investment Fluctuate So Much?," Manuscript, Federal Reserve Bank of Minneapolis.

Christiano, Lawrence J. and Martin Eichenbaum (1986): "Temporal Aggregation and Structural Inference in Macroeconomics," manuscript, Carnegie-Mellon University.

Cleveland, W. and G. Tiao (1976): "Decomposition of Seasonal Time Series: A Model for the Census X-11 Program," Journal of the American Statistical Association, 71, 581-587.

Eichenbaum, Martin S. (1984): "Rational Expectations and the Smoothing Properties of Inventories of Finished Goods," Journal of Monetary Economics, 14, 71-96.

Feldstein, Martin and Lawrence Summers (1979): "Inflation and the Taxation of Capital Income in the Corporate Sector," National Tax Journal, 32, 445-470.

Garber, Peter and Robert King (1984): "Deep Structural Excavation? A Critique of Euler Equation Methods," NBER Technical Working Paper #83-14.

Ghali, Moheb (1987): "Seasonality, Aggregation, and Testing of the Production Smoothing Hypothesis," American Economic Review, 77, 464-469.

Ghysels, Eric (1987): "Cycles and Seasonals in Inventories: Another Look at Non-Stationarity and Induced Seasonality," manuscript, University of Montreal.

Goodfriend, Marvin (1986): "Information-Aggregation Bias: The Case of Consumption," manuscript, Federal Reserve Bank of Richmond.

Hall, Robert E. (1978): "Stochastic Implications of the Life Cycle-Permanent Income Hypothesis: Theory and Evidence," Journal of Political Economy, 86, 971-987.

Hansen, Lars and Kenneth J. Singleton (1983): "Stochastic Consumption, Risk Aversion, and the Temporal Behavior of Asset Returns," Journal of Political Economy, 91, 249-266.

Hinrichs, John C. and Anthony D. Eckman (1981): "Constant Dollar Manufacturing Inventories," Survey of Current Business, 61, 16-23.

Irvine, F. Owen, Jr. (1981): "Retail Inventory Investment and the Cost of Capital," American Economic Review, 71, 633-648.

Kahn, James (1986): "Inventories and the Volatility of Production," manuscript, University of Rochester.

Maccini, Louis J. and Robert J. Rossana (1984): "Joint Production, Quasi-Fixed Factors of Production, and Investment in Finished Goods Inventories," Journal of Money, Credit, and Banking, 16, 218-236.

Mankiw, N. Gregory (1981): "The Permanent Income Hypothesis and the Real Interest Rate," Economic Letters, 7, 307-311.

Miron, Jeffrey A. (1986): "Seasonality and the Life Cycle-Permanent Income Model of Consumption," Journal of Political Economy, 94, 1258-1279.

Miron, Jeffrey A. and Stephen P. Zeldes (1987): "Production, Sales, and the Change in Inventories: An Identity that Doesn't Add Up," Rodney L. White Working Paper #20-87, The Wharton School, University of Pennsylvania.

Nelson, Charles R. and Charles I. Plosser (1982): "Trends and Random Walks in Macroeconomic Time Series: Some Evidence and Implications," Journal of Monetary Economics, 10, 139-162.

Prescott, Edward C. (1986): "Theory Ahead of Business Cycle Measurement," Carnegie-Rochester Conference Series, 25, 11-44.

Reagan, Patricia and Dennis P. Sheehan (1985): "The Stylized Facts about the Behavior of Manufacturers' Inventories and Backorders over the Business Cycle: 1959-1980," Journal of Monetary Economics, 15, 217-246.

Sargent, Thomas J. (1978): "Rational Expectations, Econometric Exogeneity, and Consumption," Journal of Political Economy, 86, 673-700.

Schutte, David P. (1983): "Inventories and Sticky Prices: Note," American Economic Review, 73, 815-816.

Summers, Lawrence (1981): "Comment on 'Retail Inventory Behavior and Business Fluctuations' by Alan Blinder," Brookings Papers on Economic Activity, 2, 513-517.

Wallis, Kenneth F. (1974): "Seasonal Adjustment and the Relation between Variables," Journal of the American Statistical Association, 69, 13-32.

Ward, Michael P. (1978): "Optimal Production and Inventory Decisions: An Analysis of Firm and Industry Behavior," unpublished Ph.D. thesis, University of Chicago.

West, Kenneth D. (1983): "A Note on the Econometric Use of Constant Dollar Inventory Series," Economic Letters, 13, 337-341.

West, Kenneth D. (1986): "A Variance Bounds Test of the Linear Quadratic Inventory Model," Journal of Political Economy, 94, 374-401.

Zeldes, Stephen P. (1985): "Consumption and Liquidity Constraints: An Empirical Investigation," Rodney L. White Center Working Paper #24-85, The Wharton School, University of Pennsylvania.

CHAPTER VI

ORDER BACKLOGS AND PRODUCTION SMOOTHING[1]

Kenneth D. West
University of Wisconsin

Abstract

Empirical examination of some aggregate manufacturing data suggests that order backlogs may help explain two puzzling facts: (1) the variability of production appears to be greater than that of demand, and (2) inventories appear to be drawn down when demand is low, built up when demand is high.

1. Introduction

The production smoothing model of inventories suggests that firms hold inventories mainly to smooth production in the face of random fluctuations in demand. It is well known, however, that some stylized facts appear to be inconsistent with both the spirit and the letter of the model. One such fact is that in virtually all manufacturing industries, the variability of production is greater than that of shipments (Blanchard (1983), Blinder (1986a), West (1986)). A second fact is that inventories tend to be accumulated when demand is high and decumulated when demand is low, precisely the opposite of the pattern predicted by the production smoothing model (Blinder (1986a), Summers (1981)).

All the studies just cited assume that physical inventories are the only buffer between demand and production. Backlogs of unfilled orders, however, might also serve as buffers. They might be built up when demand is high and drawn down when demand is low. If so, studies that ignore backlogs may be misleading.

---

[1]Paper prepared for the 1987 Conference on the Economics of Inventory Management. I thank Ben Bernanke for helpful comments and discussions, Jeff Miron for providing data and the National Science Foundation for financial support.

Indeed, in the presence of backlogs, the anomalous stylized facts probably are not even directly relevant to at least some versions of the production smoothing models. As initially stated (Holt et al. (1961)) and recently generalized (Blinder (1982)), the model does not impose a nonnegativity condition on inventories. If demand is too high, orders are put on a backlog. Backlogged orders are implicitly considered negative inventories. If the model is taken literally, the implication is that empirical studies should follow Holt et al. (1961) and Belsley (1969) and use "net" inventories, i.e., physical inventories minus backlogs. If backlogs are substantial, the bias from using physical rather than net inventories may be large.

This paper considers the anomalous stylized facts for some industries where backlogs in fact are large. It assumes a model like that in Holt et al. (1961), Belsley (1969) or Blinder (1982). The model implies that the variance of production is less than the variance of new orders (rather than shipments). This is empirically true, for the data studied here. The model also implies that the net inventory stock should buffer production from demand. The stock should be decumulated when demand is high, accumulated when demand is low. This, too, holds empirically, in two senses. First, the covariance between new orders and investment in net inventories is negative. Second, a positive shock to new orders causes net inventories to be drawn down, with production rising only gradually. On the other hand, if one ignores backlogs, and examines physical inventories and shipments instead of net inventories and new orders, the usual stylized facts result. These facts are, however, irrelevant in the present production smoothing model.

Net inventories, then, appear to smooth production in the face of random fluctuations in demand. This suggests that production smoothing may indeed be a central determinant of production.

It should be emphasized, however, that this paper does not shed direct light on the determinants of physical inventories: the model used determines net inventories, with the individual levels of physical inventories and of backlogs indeterminate. This is, of course, a serious drawback in an inventory model. Moreover, common sense, as well as some formal time series evidence (Reagan and Sheehan (1985),

West (1983b)), suggest that backlogs are not simply negative inventories. Further research is required to see whether backlogs and inventories play their prescribed roles when one allows them to affect costs in distinct ways. In addition, the evidence here is qualitative in the sense that while broad time series patterns are established, a precise model is never estimated, and standard errors are never calculated. I would therefore characterize the results in this paper as preliminary and suggestive.

Section 2 describes the model and tests performed. Section 3 presents empirical results. Section 4 concludes. An appendix available on request contains some algebraic details and empirical results omitted to save space.

2. The Model and Tests

The empirical work requires data on backlogs. The Department of Commerce only collects such data for what are called "production to order" industries. The model used will therefore be one that is appropriate for such industries.

These are industries in which orders ordinarily arrive before production is completed. Storage costs for the finished product tend to be relatively large and the product line fairly heterogeneous (Abramovitz (1950), Zarnowitz (1973)). According to Belsley (1969), most two digit industries produce primarily to order, including virtually all durable goods industries. Backlogs tend to be substantial, relative either to shipments or to physical inventories. This is illustrated for aggregate durables in Figure 1, which plots backlogs, shipments, and two measures of inventories, finished goods and the sum of finished goods and works in progress. The backlog to shipment ratio, or the (backlog - physical inventories) to shipment ratio, suggest that customers typically wait anywhere from one to five months for shipment.

Let Q_t be production, I_t physical inventories, S_t shipments, B_t backlogs (unfilled orders) and N_t new orders. The variables are linked by the identities

$$Q_t = S_t + \Delta I_t,$$

$$N_t = S_t + \Delta B_t,$$

implying

$$Q_t = N_t + \Delta H_t, \tag{1}$$

where

$$H_t \equiv I_t - B_t.$$

H_t is the net inventory stock, physical inventories minus unfilled orders.

The model I will use, which is developed in detail in the appendix, is a slightly modified version of the one in Belsley (1969). The representative firm minimizes the expected present discounted value of costs,

$$\min E_0 \sum_{t=0}^{\infty} b^t C_t \tag{2}$$

E_0 is expectations conditional on the firm's period zero information, b is the discount rate, $0<b<1$. Apart from inessential constant and linear terms, per period costs C_t are

$$C_t = a_0(\Delta Q_t + u_{1t})^2 + a_1(Q_t + u_{2t})^2 + a_2(-H_t - a_3 Q_t + u_{3t})^2. \tag{3}$$

The u_{it} are zero mean, white noise cost shocks. Apart, perhaps, from these shocks, the first two terms are standard. The cost of changing production, $a_0(\Delta Q_t + u_{1t})^2$, represents, for example, hiring and firing costs. The production cost, $a_1(Q_t + u_{2t})^2$, can be considered a Taylor series approximation to a convex cost function.

The final term in (3), $a_2(-H_t - a_3 Q_t + u_{3t})^2$, is peculiar to a production to order firm. It balances two costs. The first is a cost of having a lengthy delivery period (bad customer relations, loss of reputation, etc.). Given the rate of production Q_t, this cost increases with $-H_t$ (=backlogs-physical inventories): the bigger the backlog or the smaller the stock of physical inventories, the lengthier the delivery period. The second is a cost of having to rush production (inefficient scheduling of batch production runs, etc.) Given Q_t, this cost decreases with $-H_t$: the bigger the backlog or the smaller the stock of physical inventories, the greater the flexibility in scheduling production. See Holt et al. (1961), Childs (1967) and Belsley (1969) for further discussion. It should be noted that all the tests in this paper are robust to the possibility that $a_3=0$, in which case the model is similar to that in Blinder (1982).

I will consider two empirical implications of the model. The first concerns production variability. The model implies that net inventories are used to buffer new orders. If variables are stationary around trend, this suggests

$$0 \le var(N) - var(Q), \qquad (4)$$

where "var" is an unconditional variance. Inequality (4) follows under a variety of assumptions about market structure and demand, as long as any effects of net inventories on demand are captured by the last term in (3). In particular, (4) is implied even if prices adjust in response to demand fluctuations. See West (1986) and the appendix for a precise argument.[2]

If the variables are not stationary, var(N) and var(Q) do not exist. Related literature suggests that empirical tests that nonetheless assume that they exist may be seriously misleading (Fuller (1976), Marsh and Merton (1986)). By continuity, this also may be true in a given finite sample, if the variables are nearly nonstationary. The data used here in fact appear to be nonstationary or nearly so, even after growth is removed.

Even if the data have unit roots, ΔH_t is stationary. Since $Q_t = N_t + \Delta H_t$, N_t and Q_t are cointegrated (Engle and Granger (1987)), and a slightly more cumbersone restatement of (4) is valid. We have

---

[2] Technically, this requires $a_3=0$ and no cost shocks. If, say, the penalty for having a large backlog is prohibitive, demand shocks may be passed directly to production. In addition, if costs vary stochastically, the firm will tend to produce a relatively large amount when costs are low, thereby inducing extra variability in production. The spirit of the model, however, is that the primary role of net inventories is to buffer production from demand. It therefore seems reasonable to expect (4) to hold, even if $a_3 \ne 0$ and there are cost shocks.

$Q_t = N_t + \Delta H_t$, so $N_t^2 - Q_t^2 = -2N_t \Delta H_t - \Delta H_t^2$. Let "cov" denote an unconditional covariance. Under fairly general statistical conditions, $\text{cov}(N_t, \Delta H_t)$ exists, even if N_t has a unit root (e.g., if $(\Delta N_t, \Delta H_t)$ follows a finite parameter ARMA process; see Fuller (1976) and West (1987)). Whether or not there are unit roots, then, one can test

$$0 \le -2\text{cov}(N_t, \Delta H_t) - \text{var}(\Delta H_t). \tag{5}$$

If there are unit roots, one must not estimate $\text{cov}(N_t, \Delta H_t)$ as a sample moment in the usual way. This would just reduce (5) to (4). Section 2A explains how to get an estimate that (a) is consistent if N_t has a unit root, and (b) is asymptotically the same as (4) if the data are stationary.

The second of the model's empirical implications that I will consider concerns whether net inventories buffer production. One test of this is whether the covariance between new orders and investment in net inventories is negative (Blinder (1986a)). If so, inventories tend to be decumulated when demand is high, accumulated when demand is low. Note, however, that $\text{cov}(N_t, \Delta H_t) < 0$ is necessary (but not sufficient) for (4) and (5). Since, as we shall see, (4) and (5) hold in these data, no separate empirical work will be needed to test the proposition.

A second test of whether net inventories buffer production concerns the response of production and net inventories to a shock to new orders (Blinder (1986a)). This is conveniently analyzed under the (over) simplifying assumptions that the firm uses just lagged new orders to forecast future new orders, and that the univariate new order process follows an AR(q):

$$N_t = \phi_1 N_{t-1} + \ldots + \phi_q N_{t-q} + v_t. \tag{6}$$

In (6), unit roots are allowed (e.g., if q=1, $N_t = N_{t-1} + v_t$ is allowed). Deterministic terms are suppressed in (6) and below, for notational simplicity.

By algebra such as in Blanchard (1983) or Eichenbaum (1984), (2) and (6) imply that the decision rule for H_t is

$$H_t = \rho_1 H_{t-1} + \rho_2 H_{t-2} + \delta_0 N_t + \ldots + \delta_{q-1} N_{t-q+1} + u_t \qquad (7)$$

The disturbance u_t is a linear combination of the cost shocks u_{it}, $i=1$ to 3. The ρ_i depend on b and the a_i in a complicated way, the δ_i depend on b, and a_i and the ϕ_i in a complicated way. The exact formulas are not of interest, except perhaps to note that ρ_2 is zero if the cost of changing production a_0 is zero. Parameter estimates are consistent even if the variables have unit roots (Sims, Stock and Watson (1986)).

Under the identifying assumption that the demand shock v_t and the cost shock u_t are uncorrelated, one can estimate not only (6) but (7) as well by least squares. One can then trace out an impulse response function, for how production and net inventories respond to a demand shock v_t: $\partial H_t/\partial v_t = \delta_0$, $\partial Q_t/\partial v_t = 1+\delta_0$, $\partial H_{t+1}/\partial v_t = \rho_1\delta_0 + \delta_0\phi_1 + \delta_1$, etc. The model suggests that H_t will be drawn down in response to a positive demand shock ($\delta_0 < 0$), with production rising gradually to meet the increased demand.

3. Empirical Results
A. Data

The data were monthly and seasonally adjusted, 1967-1984. (Data that are not seasonally adjusted might be preferable (Miron and Zeldes (1986)) but are not available for backlogs.) Nominal backlog data were conveniently available from CITIBASE for aggregate durables and six two digit manufacturing industries: stone, clay and glass (SIC 32), primary metals (SIC 33), fabricated metals (SIC 34), non-electrical machinery (SIC 35), electrical machinery (SIC 36), transportation equipment (SIC 37), and instruments (SIC 38). BEA constant (1972) dollar inventory data on finished goods and work in progress inventories and shipments were kindly supplied by Jeff Miron. Inventory data were converted from cost to market as in West (1983a) and Blinder and Holtz-Eakin (1983).

Constant dollar backlog data were not available. The discussion in Foss et al. (1980, pp 156-57), as well as a reading of Bureau of the Consesus's Form M-3 (Appendix I in Foss et al. (1980)) suggests that it is reasonable to assume that firms value the entire backlog at current delivery prices. Real backlogs were therefore obtained by deflating the BEA figure for the nominal stock of backlogs by the ratio of (nominal shipments/real shipments). New orders were calculated from the identity $N_t=S_t+\Delta B_t$. Two net inventory series were used: finished goods - backlog, and finished goods + works in progress - backlog. Production was calculated as $Q_t=N_t+\Delta H_t$. As a check on the deflation procedure, real backlogs were also obtained for aggregate durables by deflating by the producer price index. The resulting second moments of the data were very similar to those reported in Table 1 below.

Before any estimation, a common geometric trend was removed from all variables. (This is consistent with the model, as shown in the appendix.) The estimated common growth rates for finished goods inventories, backlogs and shipments, in percent per month, for aggregate durables and SIC codes 32 to 38 were: .18, -.01, -.00, -.03, .29, .38, .04, .40. The estimated rates for finished goods + works in progress, backlogs and shipments were: .17, -.01, .01, -.04, .28, .40, .07, .42. Before any of the computations reported below were done, all variables were scaled to remove this growth. For example, all durables data were divided by $(1.0018)^t$ when net inventories = finished goods inventories - backlogs, by $(1.0017)^t$ when net inventories = finished goods inventories + works in progress - backlogs. Variances and covariances of the resulting data were calculated around a constant mean. Constant terms were used in estimation of (6) and (7). To make sure that inference was not sensitive to the exact estimate of growth rates, the second moments reported in Table 1 below were recalculated for aggregate durables, with growth rates half again as big or half as small (i.e., for growth rates of .17 ± (.17/2) and .18 ± (.18/2)). Results were similar.

The Durbin-Watson of each of the regressions to estimate a common trend was very low, typically under .10. This suggests possible nonstationarity of the geometrically detrended variables. To guard against possible resulting biases, the $cov(N,\Delta H)$ term that appears in equation (5) was calculated as follows. Let T be the sample size. Ignore constant terms for notational simplicity. If N_t has a unit root, $T^{-1}\Sigma N_t\Delta H_t$ has a nondegenerate limiting distribution, and thus is not a consistent estimate of $cov(N_t,\Delta H_t)$ (Fuller (1976), West (1987)).

We have $N_t = \Delta N_t + \Delta N_{t-1} + \Delta N_{t-2} + \ldots$. This suggests calculating $\text{cov}(N_t, \Delta H_t)$ as $\text{cov}(\Delta N_t, \Delta H_t) + \text{cov}(\Delta N_{t-1}, \Delta H_t) + \ldots$. Let $\hat{c}_j$ be an estimate of $\text{cov}(\Delta N_{t-j}, \Delta H_t)$, $\hat{c}_j = T^{-1}\Sigma_{t=j+1}^{T} \Delta N_{t-j} \Delta H_t$. Consider estimating $\text{cov}(N_t, \Delta H_t)$ as $\Sigma_{j=0}^{m} \hat{c}_j$, and letting $m \to \infty$ as $T \to \infty$. The literature on estimation of spectral densities (Hannan (1970, p. 280)) indicates that if $(m/T^{1/2}) \to 0$ as $m, T \to \infty$, $\Sigma_{j=0}^{m} \hat{c}_j$ consistently estimates $\text{cov}(N, \Delta H)$. I set m=20 in the results reported below. (If N_t is stationary, one could of course set m=T, and just calculate $T^{-1}\Sigma N_t \Delta H_t$.)

In equations (6) and (7) the length of the autoregression was set to four. Note that in a rational expectations environment, (7) can be estimated by OLS only if net inventories do not Granger cause new orders (i.e., only if firms use only lagged new orders to forecast future new orders). A comment in Blinder (1986a) suggests that this is roughly consistent with the data.

B. Empirical Results

Table 1 contains point estimates of the right hand sides of (4) and (5) when net inventories = finished goods inventories - backlogs, Table 2 when net inventories = finished goods + works in progress-backlogs. Units are billions of 1972 dollars, squared. As may be seen, the production variance is less than the new order variance, in all specifications except instruments (columns (4) and (6)).[3] As in Blinder (1986a), however, the production variance is almost always greater than the shipment variance (columns (5) and (7)).

Since column (4) is less than one and column (6) is positive, it follows that $\text{cov}(N_t, \Delta H_t) < 0$. Net inventories therefore on average are accumulated during expansions, decumulated during contractions. This is illustrated in Figure 2, which plots detrended aggregate durables data, for net inventories = finished goods + works in progress-backlog. The tendency for H to be built up when N is low, to be drawn down when N is high, is quite apparent. The plots of B and work in progress + finished goods inventories indicate that the theoretically predicted pattern of fluctuations for H essentially reflects procyclical accumulation of backlogs but not countercyclical accumulation of physical inventories. It is worth noting that while

---

[3] The only reason the entries for var(N) and var(S) are different in the two tables is the slightly different estimates of growth rates.

the model does not formally determine a level of inventories separate from that of backlogs, the actual inventory behavior probably is consistent with production smoothing behavior in production to order industries. Abramowitz (1950) and Belsley (1969) suggest that finished goods inventories, at least, are built up in part because of unavoidable delays in transit. One might therefore expect inventories to be built up when shipments are high.

Table 1

Second Moments, H = Finished goods - Backlogs

Industry	(1) var(Q)	(2) var(N)	(3) var(S)	(4) var(Q)/var(N)	(5) var(Q)/var(S)	(6) -2cov(N,ΔH) - cov(ΔH)	(7) -2cov(S,ΔI) - cov(ΔI)
Aggregate	4.709	8.856	4.540	.53	1.04	6.875	-.156
Stone, Clay Glasss	.060	.062	.059	.96	1.02	.006	-.002
Primary Metals	.359	.525	.366	.68	.98	.303	.017
Fabricated Metals	.167	.272	.160	.62	1.04	.196	-.012
Non-electrical Machinery	.177	.375	.161	.47	1.10	.372	-.021
Electrical Machinery	.081	.143	.076	.57	1.06	.061	-.008
Transportation Equipment	.880	2.161	.866	.41	1.02	1.305	-.011
Instruments	.006	.008	.006	.78	1.08	.000	-.001

Table 2

Second Moments, H = Finished goods + WIP - Backlogs

Industry	(1) var(Q)	(2) var(N)	(3) var(S)	(4) var(Q)/var(N)	(5) var(Q)/var(S)	(6) -2cov(N,ΔH) - cov(ΔH)	(7) -2cov(S,ΔI) - cov(ΔI)
Aggregate	5.604	8.940	4.585	.63	1.22	5.631	-1.417
Stone, Clay Glass	.060	.062	.058	.97	1.03	.006	-.002
Primary Metals	.375	.525	.370	.72	1.01	.280	.008
Fabricated Metals	.193	.279	.168	.69	1.15	.164	-.044
Non-electrical Machinery	.232	.388	.172	.60	1.35	.290	-.108
Electrical Machinery	.094	.128	.070	.74	1.35	.049	-.034
Transportation Equipment	.904	1.983	.791	.46	1.14	1.115	-.146
Instruments	.008	.007	.005	1.11	1.57	-.000	-.004

In Tables 1 and 2, columns (6) and (7) essentially calculate var(N)-var(Q) and var(S)-var(Q) in a fashion that is robust to the presence of unit roots. See the text.

Table 3

Response to Unit Demand Shock, H = Finished goods + WIP - Backlogs

	Durables				Stone, Clay and Glass		
Period	N	Q	H	Period	N	Q	H
0	1.00	.36	-.64	0	1.00	.58	-.42
1	.65	.39	-.90	1	.65	.50	-.55
12	.42	.40	-1.92	12	.47	.47	-.49
24	.19	.20	-1.91	24	.34	.35	-.41
60	.02	.04	-1.07	60	.13	.13	-.28
120	.00	.01	-.29	120	.03	.03	-.22

	Primary Metals				Fabricated Metals		
Period	N	Q	H	Period	N	Q	H
0	1.00	.25	-.75	0	1.00	.38	-.62
1	.89	.37	-1.26	1	.51	.43	-.70
12	.36	.42	-1.74	12	.38	.27	-1.98
24	.13	.18	-1.03	24	.23	.13	-3.24
60	.01	.02	-.18	60	.06	-.09	-7.36
120	.00	.00	-.01	120	.00	-.37	-21.49

	Machinery				Electrical Machinery		
Period	N	Q	H	Period	N	Q	H
0	1.00	.19	-.81	0	1.00	.25	-.75
1	.15	.16	-.80	1	.40	.23	-.91
12	.25	.24	-1.27	12	.28	.28	-1.16
24	.15	.15	-1.31	24	.15	.15	-1.23
60	.03	.04	-.98	60	.02	.02	-1.28
120	.00	.01	-.43	120	.00	.00	-1.28

	Transportation				Instruments		
Period	N	Q	H	Period	N	Q	H
0	1.00	.20	-.80	0	1.00	.44	-.56
1	.38	.25	-.92	1	.25	.29	-.53
12	.16	.13	-1.49	12	.29	.30	-.40
24	.05	.04	-1.74	24	.17	.18	-.30
60	.00	-.00	-2.14	60	.04	.04	-.16
120	.00	-.00	-2.80	120	.00	.00	-.07

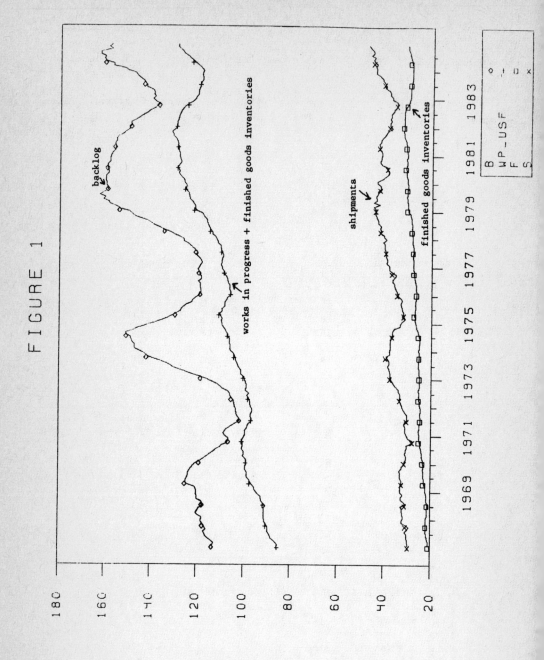

FIGURE 1

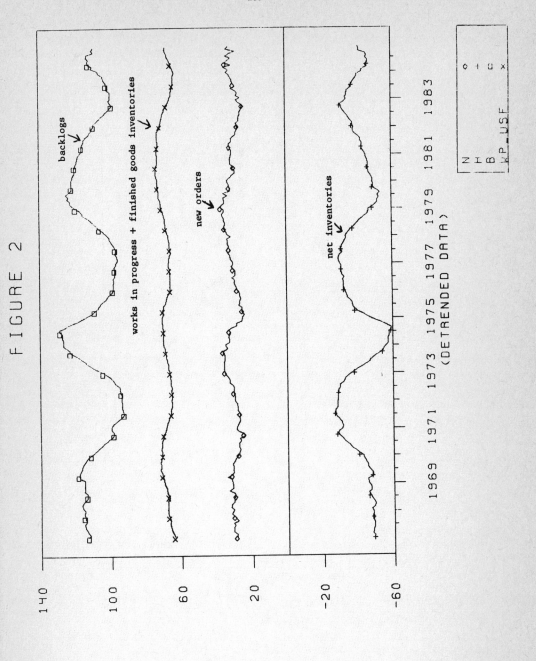

FIGURE 2

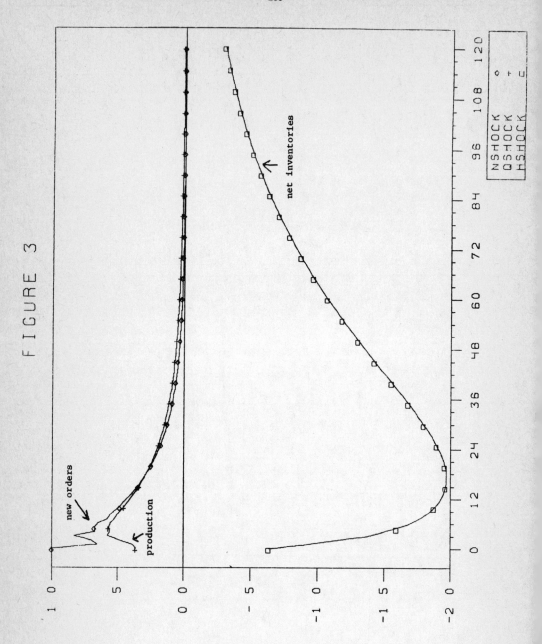

FIGURE 3

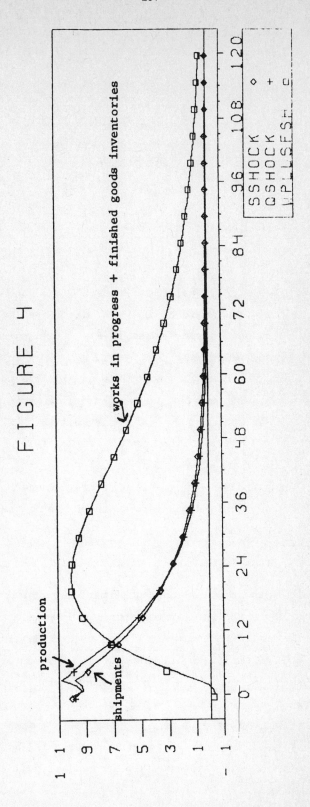

Additional evidence on the role of net inventories in buffering production may be found in the impulse response functions in Table 3. The functions are calculated from estimates of equations (6) and (7). (These estimates are available on request. Regression estimates and impulse response functions were also calculated for net inventories = finished good - backlogs, but are not reported because they were quite similar to those in Table 3.) Since the period is a month, the entry for period 12 indicates the response one year after the shock, for 24 two years after, and so on.

The estimates indicate that from 40 to 80 percent of the initial impact of a demand shock is absorbed by net inventories, with production adjusting gradually. Figure 3 contains a plot for the aggregate durables entry in Table 3. Production is built up gradually to meet the increased demand. If the data are stationary, all variables return to their steady state levels, with production meeting the increased demand ($\Sigma_{j=0}^{\infty}(\partial Q_{t+j}/\partial v_t) = \Sigma_{j=0}^{\infty}(\partial N_{t+j}/\partial v_t)$; v_t is the demand shock.) Note, however, that the return is painfully slow, indicating the borderline nonstationary behavior of inventories and new orders. In fact, the roots of $(1-\rho_1 L-\rho_2 L^2)$, with ρ_1 and ρ_2 defined in equation (6), were outside the unit circle for two data sets (fabricated metals and transportation).

Figure 4 contains the comparable plot for a shock to shipments, when physical inventories alone are assumed to buffer production. Little buffering is evident.

4. Conclusions

A production smoothing model is qualitatively consistent with some aggregate data when it is assumed that net inventories (physical inventories minus backlogs), rather than physical inventories, buffer production. The variance of production is less than that of new orders, so production is smoother than demand. The covariance of new orders and investment in net inventories is negative, so that net inventories are accumulated during contractions, decumulated during expansions. A positive shock to new orders is buffered by net inventories, so that production rises only gradually to meet increases in demand.

These results are in no sense definitive. The model that I used assumed rather implausibly that backlogs are negative inventories. No standard errors were calculated in any of the tests. The data were purely for production to order industries.

One therefore cannot jump to the conclusion that production smoothing is the major determinant of backlogs and inventories. Nonetheless, in conjunction with the conclusions of other papers, the present results seem highly suggestive. Theoretical work using more carefully formulated models than mine indicates that the presence of backlogs may indeed explain apparently anomalous production behavior (Kahn (1986), Maccini (1973)). Empirical work at least since Lovell's (1961) seminal research has found an important role for backlogs; recent contributions include Blinder (1986b) and Maccini and Rossana (1984). Large and volatile backlogs are perhaps more pervasive than many researchers, including myself (West (1986)) have assumed: of the six two digit manufacturing industries classified by Belsley (1969) as production to stock, two (apparel [SIC 23] and chemicals [SIC 28]) in fact are or have become largely production to order (Foss et al. (1980, pp. 158)).

The fundamental question is whether firms systematically use backlogs as a buffer between production and demand. If so, it is premature to conclude from, say, a comparison of production and shipment variances that firms do not smooth production in the face of fluctuations in demand. Whether or not backlogs can save the production smoothing model is therefore an important task for future research.

Appendix

This appendix deals with: (A) the derivation of (4) and (5) without growth; (B) the derivation of (4) and (5) accounting for growth; and (C) the derivation of (7).

(A) Let p_t be the real price of output. The firm maximizes

$$\max E_0 \sum_{t=0}^{\infty} b^t (p_t N_t - C_t) \tag{A1}$$

For simplicity it is assumed that the firms get revenue when an order is placed, rather than when it is shipped.

Let C_t be as in (3), with $a_3 = u_{0t} = u_{1t} = u_{2t} = 0$. Assume that all variables have zero mean. (See West (1986) for why this is an innocuous assumption.) Let V_0^* be the value of (A1), under the optimal policy. Consider an alternative policy in which $Q_t^A = N_t$ and $H_t^A = 0$ for all t. Costs C_t^A thus are $a_0 (\Delta N_t)^2 + a_1 (N_t)^2$. Under the assumption that all effects of net backlogs on demand are adequately captured by the cost function, it will still be feasible to obtain revenue $p_t N_t$ under this alternative policy. Let V_t^A be the expected present discounted value of cash flows under this alternative,

$$V_0^A = E_0 \sum_{t=0}^{\infty} b^t (p_t N_t - C_t^A).$$

Since the policy actually followed is assumed optimal, we have

$$V_0^A \leq V_0^* \quad \text{or} \quad E_0 \sum_{t=0}^{\infty} b^t (C_t^A - C_t) \geq 0 \quad \text{or}$$

$$E_0 \sum_{t=0}^{\infty} b^t \{a_0 (\Delta N_t^2 - \Delta Q_t^2) + a_1 (N_t^2 - Q_t^2) - a_2 H_t^2\} \geq 0 \tag{A2}$$

The third term is nonnegative by construction, since it is the expectation of a sum of nonnegative random variables. A necessary condition for (A2), then, is that

$$E_0 \sum_{t=0}^{\infty} b^t \{a_0(\Delta N_t^2 - \Delta Q_t^2) + a_1(N_t^2 - Q_t^2)\} \geq 0 \text{ or}$$

$$E \sum_{t=0}^{\infty} b^t \{a_0(\Delta N_t^2 - \Delta Q_t^2) + a_1(N_t^2 - Q_t^2)\} \geq 0$$

$$= (1-b)^{-1} \{a_0[\text{var}(\Delta N) - \text{var}(\Delta Q)] + a_1 E(N_t^2 - Q_t^2)\} \geq 0 \qquad (A3)$$

The first implication follows from the law of iterated expectations, since, under the assumption that at most one difference is required to induce stationarity, each term in each infinite sum has a finite unconditional expectation.

In versions of the model in which $a_0=0$ (e.g., Blinder (1986a)), equations (4) and (5) follow from (A3). If $a_0 \neq 0$ (e.g., Belsley (1969)), (4) and (5) follow if $\text{var}(\Delta N) - \text{var}(\Delta Q) > 0$. Although not reported in the paper, the sample variances of ΔN and ΔQ obeyed this inequality for all eight data sets.

(B) As stated in the text, the data actually used in the regressions were scaled by a growth rate of $(1+g)^t$. It is assumed that $b(1+g)<1$. In explaining how this fits into the model, it is convenient to call h_t, q_t and n_t the original data in levels and H_t, Q_t, and N_t the scaled data (e.g., $H_t = h_t/(1+g)^t$). The full cost function, including deterministic terms, is

$$C_t = k_t + c_{0t}(\Delta q_t - m_{0t}) + c_{1t}(q_t - m_{1t}) + c_{2t}(-h_t - a_3 q_t - m_{2t})$$
$$+ a_0(\Delta q_t - m_{0t})^2 + a_1(q_t - m_{1t})^2 + a_2(-h_t - a_3 q_t - m_{2t})^2 \qquad (A4)$$

where k_t is a purely deterministic term that grows no faster than $(1+g)^t$. The m_{it} shift the minimum cost points for each of the three types of costs. Each m_{it} has both deterministic and stochastic components, $m_{it} = (1+g)^t m_i - (1+g)^t u_{it}$. The u_{it} are the white noise cost shocks in equation (2). The c_{it} grow deterministically, $c_{it} = (1+g)^t c_i$.

respect to h_t, setting the resulting expression equal to zero, dividing by 2, and rearranging gives

$$E_t\{b^2 a_0 h_{t+2} - [a_0(2b+2b^2)+ba_1+ba_2 a_3(1+a_3)]h_{t+1}$$

$$+ [a_0(1+4b+b^2)+a_1(1+b)+a_2(1+a_3)^2+ba_2 a_3^2]h_t$$

$$- [a_0(2+2b)+a_1+a_2 a_3(1+a_3)]h_{t-1} + a_0 h_{t-2}$$

$$+ b^2 a_0 n_{t+2} - [a_0 * b^2+2b)+ba_1+ba_2 a_3^2]n_{t+1}$$

$$+ [a_0(1+2b)+a_1 a_2 a_3(1+a_3)]n_t - a_0 n_{t-1}$$

$$+ (1+g)^t c + (1+g)^t m - (1+g)^t f_t\} = 0, \qquad (A5)$$

where

$$c = c_0(1-2b+b^2) + c_1(1-b) + c_2[-(1+a_3)+ba_3],$$

$$m = a_0 m_0[1-2b(1+g)+b^2(1+g)^2] + a_1 m_1[1-b(1+g)]$$

$$+ a_2 m_2[-(1+a_3)+ba_3(1+g)],$$

$$v_t = a_0 u_{0t} + a_1 u_{1t} - a_2(1+a_3)u_{2t}.$$

Dividing by $(1+g)^t$ and rearranging terms gives

$$E_t\{[(1+g)b]^2 a_0 H_{t+2} - (1+g)[a_0(2b+2b^2)+ba_1+ba_2 a_3(1+a_3)]H_{t+1}$$

$$+ [a_0(1+4b+b^2)+a_1(1+b)+a_2(1+a_3)^2+ba_2 a_3^2]H_t$$

$$- (1+g)^{-1}[a_0(2+2b)+a_1+a_2 a_3(1+a_3)]H_{t-1} + (1+g)^{-2} a_0 H_{t-2} = Z_{t+2}\},$$
$$(A6)$$

where

$$Z_{t+2} = - [(1+g)b]^2 a_0 N_{t+2} + (1+g)[a_0(b^2+2b)+ba_1+ba_2 a_3^2] N_{t+1}$$

$$- [a_0(1+2b)+a_1+a_2 a_3(1+a_3)] N_t + (1+g)^{-1} a_0 N_{t-1}$$

$$- c - m + v_t \tag{A6}$$

It follows in part (C) below that if N_t is stationary in levels or some diference, then so is H_t. Thus, if N_t grows at rate $1+g$, so does H_t.

(C) Call λ_1 and λ_2 the two smallest (in modulus) roots to the fourth degree (A5) lag polynomial in H_t. The comparable polynomial in (A6) has roots $\lambda_1/(1+g)$, $\lambda_2/(1+g)$, $1/[b(1+g)\lambda_1]$ and $1/[b(1+g)\lambda_2]$. So if $|\lambda_1|<1$ and $|\lambda_2|<1$, the (A6) lag polynomail has exactly two stable and two unstable roots. Let $\rho_1=-(\lambda_1+\lambda_2)/(1+g)$, $\rho_2=(\lambda_1\lambda_2)/(1+g)^2$. Solving the stable roots backwards and the unstable roots forwards gives

$$H_t = \rho_1 H_{t-1} + \rho_2 H_{t-2} + dE_t \{b(1+g)\lambda_1 \sum_{j=0}^{\infty} [b(1+g)\lambda_1]^j Z_{t+j+2}$$

$$- b(1+g)\lambda_2 \sum_{j=0}^{\infty} [b(1+g)\lambda_2]^j Z_{t+j+2} \}, \tag{A7}$$

where

$$d = \lambda_1 \lambda_2 / [(\lambda_1 - \lambda_2) b(1+g) a_0].$$

It follows from the argument in Blanchard (1983) that if the firm uses only past new orders to forecast future new orders, and that equation (6) represents the univariate new orders process, then equation (7) is the closed form solution to (A7).

References

Abramovits, Moses (1950): <u>Inventories and Business Cycles, with Special Reference to Manufacturer's Inventories</u>, New York: NBER.

Belsley, David A. (1969): <u>Industry Production Behavior: The Order-Stock Distinction</u>, Amsterdam: North-Holland.

Blinder, Alan S. (1982): "Inventories and Sticky Prices: More on the Microfoundations of Microeconomics," <u>American Economic Review</u> 72, 334-48.

Blinder, Alan S. (1986a): "Can the Production Smoothing Model of Inventory Behavior be Saved?," <u>The Quarterly Journal of Economics</u> CI, 431-454.

Blinder, Alan S. (1986b): "More on the Speed of Adjustment in Inventory Models," <u>Journal of Money, Credit and Banking</u>, 355-65.

Blinder, Alan S. and Douglas Holtz-Eakin (1983): "Constant Dollar Manufacturer's Inventories of Finished Goods," <u>Journal of Monetary Economics</u> 14, 71-96.

Childs, Gerald D. (1967): <u>Inventories and Unfilled Orders</u>, Amsterdam: North Holland.

Eichenbaum, Martin S. (1984): "Rational Expectations and the Smoothing Properties of Inventories of Finished Goods," <u>Journal of Monetary Economics</u> 14, 701-96.

Engle, Robert F., and C.W.J. Granger (1987): "Dynamic Model Specification with Equilibrium Constraints: Co-integration and Error Correction," <u>Econometrica</u> 55, 251-276.

Foss, Murray F., Fromm, Gary, and Irving Rottenberg (1980): <u>Measurement of Business Inventories</u>, Washington: Government Printing Office.

Fuller, Wayne A. (1976): <u>Introduction to Statistical Time Series</u>, New York: John Wiley and Sons.

Holt, Charles C., Modigliani, Franco, Muth John, and Herbert Simon (1960): <u>Planning Production, Inventories and Work Force</u>, Englewood Cliffs, N.J.: Prentice Hall.

Kahn, James A. (1986): "Inventories and the Volatility of Production," University of Rochester Department of Economics Working Paper No. 66.

Lovell, Michael C. (1961): "Manufacturers' Inventories, Sales Expectations, and the Accelerator Principle," <u>Econometrica</u> 29, 293-314.

Maccini, Louis M. (1973): "On Optimal Delivery Lags," *Journal of Economic Theory* 6, 107-125.

Maccini, Louis M., and Robert Rossana (1984): "Joint Production, Quasi-Fixed Factors of Production and Investment in Finished Goods Inventories," *Journal of Money, Credit and Banking* 16, 218-236.

Marsh, Terry A. and Robert C. Merton (1986): "Dividend Variability and Variance Bounds Test for the Rationality of Stock Market Prices," *American Economic Review* 76, 483-398.

Miron, Jeffrey A. and Stephen P. Zeldes (1986): "Seasonality, Cost Shocks and the Production Smoothing Model of Inventories," manuscript, University of Pennsylvania.

Reagan, Patricia and Dennis P. Sheahan (1985): "The Stylized Facts about the Behavior of Manufacturers' Backorders and Inventories over the Business Cycle, 1959-1980," *Journal of Monetary Economics* 15, 217-246.

Sims, Christopher A., Stock, James H., and Mark W. Watson (1986): "Inference in Linear time Series with Some Unit Roots," manuscript, University of Minnesota.

Summers, Lawrence (1981): "Comment," *Brookings Papers on Economic Activity*, 513-517.

West, Kenneth D. (1983a): "A Note on the Econometric Use of Constant Dollar Inventory Series," *Economics Letters* 13, 337-341.

West, Kenneth D. (1983b): "Inventory Models and Backlog Costs: An Empirical Investigation," Massachusetts Institute of Technology Ph.D. thesis.

West, Kenneth D. (1986): "A Varianace Bounds Test of the Linear Quadratic Inventory Model," *Journal of Political Economy* 94, 374-401.

West, Kenneth D. (1987): "Asymptotic Normality, When Regressors Have a Unit Root," Princeton University Woodrow Wilson School Discussion Paper No. 110, revised.

Zarnowitz, Victor (1973): *Orders, Production and Investment--A Cyclical and Structural Analysis*, New York: Columbia University Press.

Vol. 236: G. Gandolfo, P.C. Padoan, A Disequilibrium Model of Real and Financial Accumulation in an Open Economy. VI, 172 pages. 1984.

Vol. 237: Misspecification Analysis. Proceedings, 1983. Edited by T. K. Dijkstra. V, 129 pages. 1984.

Vol. 238: W. Domschke, A. Drexl, Location and Layout Planning. IV, 134 pages. 1985.

Vol. 239: Microeconomic Models of Housing Markets. Edited by K. Stahl. VII, 197 pages. 1985.

Vol. 240: Contributions to Operations Research. Proceedings, 1984. Edited by K. Neumann and D. Pallaschke. V, 190 pages. 1985.

Vol. 241: U. Wittmann, Das Konzept rationaler Preiserwartungen. XI, 310 Seiten. 1985.

Vol. 242: Decision Making with Multiple Objectives. Proceedings, 1984. Edited by Y. Y. Haimes and V. Chankong. XI, 571 pages. 1985.

Vol. 243: Integer Programming and Related Areas. A Classified Bibliography 1981–1984. Edited by R. von Randow. XX, 386 pages. 1985.

Vol. 244: Advances in Equilibrium Theory. Proceedings, 1984. Edited by C.D. Aliprantis, O. Burkinshaw and N.J. Rothman. II, 235 pages. 1985.

Vol. 245: J.E.M. Wilhelm, Arbitrage Theory. VII, 114 pages. 1985.

Vol. 246: P.W. Otter, Dynamic Feature Space Modelling, Filtering and Self-Tuning Control of Stochastic Systems. XIV, 177 pages. 1985.

Vol. 247: Optimization and Discrete Choice in Urban Systems. Proceedings, 1983. Edited by B.G. Hutchinson, P. Nijkamp and M. Batty. VI, 371 pages. 1985.

Vol. 248: Plural Rationality and Interactive Decision Processes. Proceedings, 1984. Edited by M. Grauer, M. Thompson and A.P. Wierzbicki. VI, 354 pages. 1985.

Vol. 249: Spatial Price Equilibrium: Advances in Theory, Computation and Application. Proceedings, 1984. Edited by P. T. Harker. VII, 277 pages. 1985.

Vol. 250: M. Roubens, Ph. Vincke, Preference Modelling. VIII, 94 pages. 1985.

Vol. 251: Input-Output Modeling. Proceedings, 1984. Edited by A. Smyshlyaev. VI, 261 pages. 1985.

Vol. 252: A. Birolini, On the Use of Stochastic Processes in Modeling Reliability Problems. VI, 105 pages. 1985.

Vol. 253: C. Withagen, Economic Theory and International Trade in Natural Exhaustible Resources. VI, 172 pages. 1985.

Vol. 254: S. Müller, Arbitrage Pricing of Contingent Claims. VIII, 151 pages. 1985.

Vol. 255: Nondifferentiable Optimization: Motivations and Applications. Proceedings, 1984. Edited by V.F. Demyanov and D. Pallaschke. VI, 350 pages. 1985.

Vol. 256: Convexity and Duality in Optimization. Proceedings, 1984. Edited by J. Ponstein. V, 142 pages. 1985.

Vol. 257: Dynamics of Macrosystems. Proceedings, 1984. Edited by J.-P. Aubin, D. Saari and K. Sigmund. VI, 280 pages. 1985.

Vol. 258: H. Funke, Eine allgemeine Theorie der Polypol- und Oligopolpreisbildung. III, 237 pages. 1985.

Vol. 259: Infinite Programming. Proceedings, 1984. Edited by E.J. Anderson and A.B. Philpott. XIV, 244 pages. 1985.

Vol. 260: H.-J. Kruse, Degeneracy Graphs and the Neighbourhood Problem. VIII, 128 pages. 1986.

Vol. 261: Th. R. Gulledge, Jr., N.K. Womer, The Economics of Made-to-Order Production. VI, 134 pages. 1986.

Vol. 262: H.U. Buhl, A Neo-Classical Theory of Distribution and Wealth. V, 146 pages. 1986.

Vol. 263: M. Schäfer, Resource Extraction and Market Structure. XI, 154 pages. 1986.

Vol. 264: Models of Economic Dynamics. Proceedings, 1983. Edited by H.F. Sonnenschein. VII, 212 pages. 1986.

Vol. 265: Dynamic Games and Applications in Economics. Edited by T. Başar. IX, 288 pages. 1986.

Vol. 266: Multi-Stage Production Planning and Inventory Control. Edited by S. Axsäter, Ch. Schneeweiss and E. Silver. V, 264 pages. 1986.

Vol. 267: R. Bemelmans, The Capacity Aspect of Inventories. IX, 165 pages. 1986.

Vol. 268: V. Firchau, Information Evaluation in Capital Markets. VII, 103 pages. 1986.

Vol. 269: A. Borglin, H. Keiding, Optimality in Infinite Horizon Economies. VI, 180 pages. 1986.

Vol. 270: Technological Change, Employment and Spatial Dynamics. Proceedings 1985. Edited by P. Nijkamp. VII, 466 pages. 1986.

Vol. 271: C. Hildreth, The Cowles Commission in Chicago, 1939–1955. V, 176 pages. 1986.

Vol. 272: G. Clemenz, Credit Markets with Asymmetric Information. VIII, 212 pages. 1986.

Vol. 273: Large-Scale Modelling and Interactive Decision Analysis. Proceedings, 1985. Edited by G. Fandel, M. Grauer, A. Kurzhanski and A.P. Wierzbicki. VII, 363 pages. 1986.

Vol. 274: W.K. Klein Haneveld, Duality in Stochastic Linear and Dynamic Programming. VII, 295 pages. 1986.

Vol. 275: Competition, Instability, and Nonlinear Cycles. Proceedings, 1985. Edited by W. Semmler. XII, 340 pages. 1986.

Vol. 276: M.R. Baye, D.A. Black, Consumer Behavior, Cost of Living Measures, and the Income Tax. VII, 119 pages. 1986.

Vol. 277: Studies in Austrian Capital Theory, Investment and Time. Edited by M. Faber. VI, 317 pages. 1986.

Vol. 278: W.E. Diewert, The Measurement of the Economic Benefits of Infrastructure Services. V, 202 pages. 1986.

Vol. 279: H.-J. Büttler, G. Frei and B. Schips, Estimation of Disequilibrium Models. VI, 114 pages. 1986.

Vol. 280: H.T. Lau, Combinatorial Heuristic Algorithms with FORTRAN. VII, 126 pages. 1986.

Vol. 281: Ch.-L. Hwang, M.-J. Lin, Group Decision Making under Multiple Criteria. XI, 400 pages. 1987.

Vol. 282: K. Schittkowski, More Test Examples for Nonlinear Programming Codes. V, 261 pages. 1987.

Vol. 283: G. Gabisch, H.-W. Lorenz, Business Cycle Theory. VII, 229 pages. 1987.

Vol. 284: H. Lütkepohl, Forecasting Aggregated Vector ARMA Processes. X, 323 pages. 1987.

Vol. 285: Toward Interactive and Intelligent Decision Support Systems. Volume 1. Proceedings, 1986. Edited by Y. Sawaragi, K. Inoue and H. Nakayama. XII, 445 pages. 1987.

Vol. 286: Toward Interactive and Intelligent Decision Support Systems. Volume 2. Proceedings, 1986. Edited by Y. Sawaragi, K. Inoue and H. Nakayama. XII, 450 pages. 1987.

Vol. 287: Dynamical Systems. Proceedings, 1985. Edited by A.B. Kurzhanski and K. Sigmund. VI, 215 pages. 1987.

Vol. 288: G.D. Rudebusch, The Estimation of Macroeconomic Disequilibrium Models with Regime Classification Information. VII, 128 pages. 1987.

Vol. 289: B.R. Meijboom, Planning in Decentralized Firms. X, 168 pages. 1987.

Vol. 290: D.A. Carlson, A. Haurie, Infinite Horizon Optimal Control. XI, 254 pages. 1987.

Vol. 291: N. Takahashi, Design of Adaptive Organizations. VI, 140 pages. 1987.

Vol. 292: I. Tchijov, L. Tomaszewicz (Eds.), Input-Output Modeling. Proceedings, 1985. VI, 195 pages. 1987.

Vol. 293: D. Batten, J. Casti, B. Johansson (Eds.), Economic Evolution and Structural Adjustment. Proceedings, 1985. VI, 382 pages. 1987.

Vol. 294: J. Jahn, W. Krabs (Eds.), Recent Advances and Historical Development of Vector Optimization. VII, 405 pages. 1987.

Vol. 295: H. Meister, The Purification Problem for Constrained Games with Incomplete Information. X, 127 pages. 1987.

Vol. 296: A. Börsch-Supan, Econometric Analysis of Discrete Choice. VIII, 211 pages. 1987.

Vol. 297: V. Fedorov, H. Läuter (Eds.), Model-Oriented Data Analysis. Proceedings, 1987. VI, 239 pages. 1988.

Vol. 298: S.H. Chew, Q. Zheng, Integral Global Optimization. VII, 179 pages. 1988.

Vol. 299: K. Marti, Descent Directions and Efficient Solutions in Discretely Distributed Stochastic Programs. XIV, 178 pages. 1988.

Vol. 300: U. Derigs, Programming in Networks and Graphs. XI, 315 pages. 1988.

Vol. 301: J. Kacprzyk, M. Roubens (Eds.), Non-Conventional Preference Relations in Decision Making. VII, 155 pages. 1988.

Vol. 302: H.A. Eiselt, G. Pederzoli (Eds.), Advances in Optimization and Control. Proceedings, 1986. VIII, 372 pages. 1988.

Vol. 303: F.X. Diebold, Empirical Modeling of Exchange Rate Dynamics. VII, 143 pages. 1988.

Vol. 304: A. Kurzhanski, K. Neumann, D. Pallaschke (Eds.), Optimization, Parallel Processing and Applications. Proceedings, 1987. VI, 292 pages. 1988.

Vol. 305: G.-J.C.Th. van Schijndel, Dynamic Firm and Investor Behaviour under Progressive Personal Taxation. X, 215 pages. 1988.

Vol. 306: Ch. Klein, A Static Microeconomic Model of Pure Competition. VIII, 139 pages. 1988.

Vol. 307: T.K. Dijkstra (Ed.), On Model Uncertainty and its Statistical Implications. VII, 138 pages. 1988.

Vol. 308: J.R. Daduna, A. Wren (Eds.), Computer-Aided Transit Scheduling. VIII, 339 pages. 1988.

Vol. 309: G. Ricci, K. Velupillai (Eds.), Growth Cycles and Multisectoral Economics: the Goodwin Tradition. III, 126 pages. 1988.

Vol. 310: J. Kacprzyk, M. Fedrizzi (Eds.), Combining Fuzzy Imprecision with Probabilistic Uncertainty in Decision Making. IX, 399 pages. 1988.

Vol. 311: R. Färe, Fundamentals of Production Theory. IX, 163 pages. 1988.

Vol. 312: J. Krishnakumar, Estimation of Simultaneous Equation Models with Error Components Structure. X, 357 pages. 1988.

Vol. 313: W. Jammernegg, Sequential Binary Investment Decisions. VI, 156 pages. 1988.

Vol. 314: R. Tietz, W. Albers, R. Selten (Eds.), Bounded Rational Behavior in Experimental Games and Markets. VI, 368 pages. 1988.

Vol. 315: I. Orishimo, G.J.D. Hewings, P. Nijkamp (Eds.), Information Technology: Social and Spatial Perspectives. Proceedings, 1986. VI, 268 pages. 1988.

Vol. 316: R.L. Basmann, D.J. Slottje, K. Hayes, J.D. Johnson, D.J. Molina, The Generalized Fechner-Thurstone Direct Utility Function and Some of its Uses. VIII, 159 pages. 1988.

Vol. 317: L. Bianco, A. La Bella (Eds.), Freight Transport Planning and Logistics. Proceedings, 1987. X, 568 pages. 1988.

Vol. 318: T. Doup, Simplicial Algorithms on the Simplotope. VIII, 262 pages. 1988.

Vol. 319: D.T. Luc, Theory of Vector Optimization. VIII, 173 pages. 1989.

Vol. 320: D. van der Wijst, Financial Structure in Small Business. VII, 181 pages. 1989.

Vol. 321: M. Di Matteo, R.M. Goodwin, A. Vercelli (Eds.), Technological and Social Factors in Long Term Fluctuations. Proceedings. IX, 442 pages. 1989.

Vol. 322: T. Kollintzas (Ed.), The Rational Expectations Equilibrium Inventory Model. XI, 269 pages. 1989.